PSYCHIATRY - THEORY, APPLICATIONS AND TREATMENTS

MODULATORS OF GLUTAMATERGIC SIGNALING AS POTENTIAL TREATMENTS FOR NEUROPSYCHIATRIC DISORDERS

Psychiatry - Theory, Applications and Treatments

Additional books in this series can be found on Nova's website
under the Series tab.

Additional e-books in this series can be found on Nova's website
under the e-book tab.

PSYCHIATRY - THEORY, APPLICATIONS AND TREATMENTS

MODULATORS OF GLUTAMATERGIC SIGNALING AS POTENTIAL TREATMENTS FOR NEUROPSYCHIATRIC DISORDERS

ZORAN M. PAVLOVIC
EDITOR

NOVA BIOMEDICAL

New York

Library of Congress Cataloging-in-Publication Data

ISBN: 978-1-63483-493-3

Library of Congress Control Number: 2015946400

Published by Nova Science Publishers, Inc. † New York

To my parents Milan and Nadezda, my brother Bratislav and my daughter Mariza

CONTENTS

PREFACE

Dear Reader,

As a part of introduction I would like to provide you with main reasons and personal incentives why I embarked on this project as Editor and assembled my expert colleagues to write on glutamate, its role in pathophysiology of neuropsychiatric disorders, and current and future glutamate based treatment strategies.

My interest for glutamate started about a decade ago in my psychiatric practice, after I treated several of my patients with lamotrigine and published these successful patient stories as case reports in distinguished neuropsychiatric peer reviewed journals. Therefore I would consider this book release as an extension of these past activities but also as a significant contribution to the field of neuropsychopharmacology after more than 20 years of my work in various roles within pharma industry, clinical practice and clinical research organizations.

I sincerely hope that you will enjoy its content and also get some insights that might be useful for your future research and clinical work.

As the Editor my role would be to provide you with a glimpse into the glutamate history, its specific receptors and extensive past and ongoing research activities related to glutamate involvement in major neuropsychiatric disorders, and together with my colleagues help you understand the basic mechanisms which are driving future growth in discovery of novel glutamatergic treatments.

INTRODUCTION

It is commonly known that glutamate and glutamate based therapies gained a lot of attention in the last couple of decades due to glutamate role as a major excitatory neurotransmitter in the brain, exciting ketamine "story" in treatment resistant depression, and recent discovery that several novel therapeutic agents share glutamatergic properties as their mechanism of action. Furthermore, a large number of glutamate receptor selective compounds are currently tested in clinical trials for diseases that span a wide range of CNS disorders.

Glutamate History

Glutamic acid was discovered in the 1866, by the German chemist Karl Heinrich Ritthausen and its salts are known as glutamates. In human beings glutamate is the most abundant excitatory neurotransmitter in the central nervous system, exerts its effects through the activation of several receptor subtypes and mediates fast excitatory synaptic transmission between neurons. In addition to classical receptor mediated neurotransmission glutamate also spills-over and modulates synaptic crosstalk and plays important role in the regulation of synaptogenesis. The glutamate story in neuroscience field started in the 1930s by recognizing neurophysiological role of this amino acid in the brain, and this led to first trials the 1940s in terms of dietary use of glutamate and glutamines in the treatment of learning disorders and epilepsy. In 1977 John Davies did the experiment that finally confirmed that NMDA receptors were indeed synaptic receptors and this discovery led to further investigations and availability of potent and selective NMDA receptor antagonists synthesized in various scientific laboratories in the 80s. These preclinical developments boosted clinician involvement mainly through their findings that some drugs were able to block particular glutamate receptors. First drug discovered by Lodge et al. to have effect on glutamatergic NMDA receptors was dissociative anesthetic ketamine, which selectively reduced excitation of central mammalian neurons. In the 90s further distinction between two major families of glutamate receptors was recognized, and classification to those that mediate fast synaptic responses by opening ion channels, known as ionotropic glutamate (iGlu) receptors, and those that cause slower synaptic effects changes, known as metabotropic glutamate (mGlu) receptors, was made. After we entered the new millennium the field of scientific knowledge related to glutamate really flourished and resulted in detailed description of glutamate transporters family, consisting of five Na+-dependent glutamate transporters (EAATs 1-5) and three vesicular glutamate transporters (VGLUTs 1-3). Furthermore the evolution of pharmacogenetics and identification of a genetic variability involvement in for instance response to antidepressants, reported in the Sequenced Treatment Alternatives to Relieve Depression (STAR*D) clinical trial, further emphasized the importance of glutamate in pathophysiology of major CNS disorders. Of particular importance is that in the last two decades various neurological and psychiatric therapies affecting glutamatergic neurotransmission such as topiramate for epilepsy, memantine for Alzheimer's Dementia and ketamine use in treatment resistant depression, and recently in PTSD, paved the way for more novel approaches in designing specific treatment modalities. These potential drugs "design improvements," were introduced by creation of positive and negative allosteric modulators of glutamate receptor subtypes, in order to provide "fine tuning" of central synaptic glutamatergic processes. Last but not least, new models of alliances between academic institutions and pharma industry started to take place in the last couple of years, which would hopefully speed up drug discovery process and introduce new treatments for various neuropsychiatric disorders including neurodevelopmental disorders such as autism, childhood ADHD and adolescent bipolar disorder.

Glutamate Receptors

Ionotropic Receptors

There are three distinct subtypes of ionotropic glutamate receptors (GluRs) that have been identified including 2-amino-3-(5-methyl-3-oxo-1,2-oxazol-4-yl) propanoic acid receptors (AMPARs), N-methyl-D-aspartate receptors (NMDARs) and kainate receptors. The most common GluRs in mature synapses are AMPARs that mediate the fast excitatory neurotransmission and NMDARs that mediate the slow excitatory neurotransmission.

NMDA Receptor

The N-methyl-D-aspartate receptor (also known as the NMDA receptor or NMDAR) is a specific type of ionotropic glutamate. A functional NMDAR is a heterotetramer composed mainly of two GluN1 subunits and two GluN2 subunits; in some cases the GluN3 subunits are also incorporated into the heterotetramer [1]. In the synaptic region NMDARs are found on all parts, including the pre- and postsynaptic membranes but also in extrasynaptical areas that surround the synapse. At least in some cases, activation of synaptic NMDAR pathways may upregulate neuronal cell functional plasticity and survival, while activation of extrasynaptic ones may turn on pathways leading to neurodegeneration [2]. Furthermore activation of synaptic vs. extrasynaptic NMDARs may also be tied to pathways leading to long-term potentiation (LTP) vs. long-term depression (LTD), respectively (3). During the process of development, changes occur in number and composition of NMDAR receptors. NMDARs in early postnatal development mainly contain GluN2B or GluN2D, while GluN2A and GluN2C subunits are more prevalent in adults. Stimuli that induce LTP induce the switch from GluN2B- to GluN2A-containing NMDARs in young animals (4). Also as noted above, GluN3A is prevalent in early development and is lost or reduced with age. GluN3A is believed to prevent the stabilization of premature synapses, and its downregulation is necessary for synapse maturation [5]. Generally, it has been thought that change in NMDAR composition at synapses is a developmental phenomenon accompanying the increase of AMPARs to adult levels [6], and that adult plasticity involves changes in number and composition of AMPARs but not NMDARs [7].

AMPA Receptors

The α-amino-3-hydroxy-5-methyl-4-isoxazolepropionic acid receptor (also known as AMPA receptor, AMPAR) is a non-NMDA-type ionotropic transmembrane receptor for glutamate that mediates fast synaptic transmission in the central nervous system (CNS). AMPARs are composed of four types of subunits, designated as GluR1 (GRIA1), GluR2 (GRIA2), GluR3 (GRIA3), and GluR4, alternatively called GluRA-D2 (GRIA4), which combine to form tetramers. Most AMPARs are heterotetrameric, consisting of symmetric 'dimer of dimers' of GluR2 and either GluR1, GluR3 or GluR4. AMPARs are both glutamate receptors and cation channels that are integral to plasticity and synaptic transmission at many postsynaptic membranes. The most widely and thoroughly investigated forms of plasticity in the nervous system is known as long-term potentiation, or LTP. In normal ageing process in laboratory aged animals show either a higher threshold for LTP induction or a decreased level of LTP induction compared with young animals. Thus, while aged animals still exhibit LTP, higher levels of stimulation are required and the potentiation is less stable [8].

As for long-term depression (LTD), it enacts mechanisms of decrease AMPA receptor density in selected dendritic spines, such as decreased AMPAR surface expression through increased AMPAR endocytosis.

Kainate Receptors

There are five types of kainate receptor subunits, GluR5 (GRIK1), GluR6 (GRIK2), GluR7 (GRIK3), KA1 (GRIK4) and KA2 (GRIK5), arranged in different ways to form a tetramer, a four subunit receptor. Kainate receptors have both presynaptic and postsynaptic actions but do have more limited distribution in the brain than AMPA and NMDA receptors. Unlike AMPA receptors, kainate receptors play only a minor role in signaling at synapses and have a more subtle role in synaptic plasticity, affecting the likelihood that the postsynaptic cell will fire in response to future stimulation. Their probably most unique functional feature is their ability despite operating in ionotropic signaling, to be involved in metabotropic signaling pathways as well.

Metabotropic Receptors

The metabotropic glutamate receptors, or mGluRs, are glutamate receptors, members of the group C family of G-protein-coupled receptors, or GPCRs that participate in the modulation of synaptic transmission and neuronal excitability throughout the central nervous system. GPCRs are membrane-bound proteins that are activated by extracellular ligands such as light, peptides, and neurotransmitters, and transduce intracellular signals via interactions with G proteins. Metabotropic glutamate receptors are known to act as modulators of (affect the activity of) other receptors. For example, group I mGluRs are known to increase the activity of N-methyl-D-aspartate (NMDARs). On the other hand, agonists of group II and III mGluRs reduce NMDAR activity, thus protecting neurons from excitotoxicity. Group I includes mGluRs 1 and 5, Group II includes mGluRs 2 and 3, and Group III includes mGluRs, 4, 6, 7, and 8. In general, group I mGluRs couple to G q/G11and activate phospholipase Cβ, resulting in the hydrolysis of phosphotinositides and generation of inositol 1,4,5-trisphosphate (IP3) and diacyl-glycerol. This classical pathway leads to calcium mobilization and activation of protein kinase C (PKC). In contrast to group I mGluRs, group II and III mGluRs are coupled predominantly to Gi/o proteins. Gi/o linked receptors are classically coupled to the inhibition of adenylyl cyclase and directly regulate ion channels and other downstream signaling partners via liberation of G βγ subunits. As with group I mGluRs, it is becoming increasingly appreciated that group II and group III mGluRs also couple to other signaling pathways, including activation of mitogen-activated protein kinase (MAPK) and phosphatidyl inositol 3-kinase (PI3) kinase pathways [9], providing further complexity regarding the mechanisms by which these receptors can regulate synaptic transmission.

Glutamate Transporters

There are a number of transporter families that transport glutamate, and include the membrane excitatory amino acid transporters (EAATs), the vesicular glutamate transporters (VGLUTs), and the glutamate-cysteine exchanger. As excessive glutamate receptor stimulation is toxic to neurons, role of glutamate transporters is to rapidly clear glutamate from the synapse and reduce the extent of glutamate spillover from the synapse. There are five subtypes of EAAT transporters. EAAT1 transporter is the major glutamate transporter in the cerebellum. As for EAAT2 it is responsible for 90–95% of glutamate uptake in the forebrain and is predominantly expressed in astrocytes, but also up to 10% of EAAT2 is expressed in presynaptic neuronal terminals in the hippocampus. EAAT3 is expressed in highest concentrations in the hippocampus, cerebellum, and basal ganglia, but the level of expression is 100-fold less than those of EAAT1 and EAAT2. A large proportion of EAAT3 is intracellular, but it is thought that it may be rapidly mobilized to be expressed in the membrane as well. Furthermore, EAAT3 is predominantly expressed in the dendrites and soma, which suggest that it plays a homeostatic or cell metabolism role rather than in directly regulating synaptic transmission. EAAT4 expression is highly localized, with most EAAT4 expressed in the Purkinje cells of the cerebellum and also with low levels of expression in other regions such as the forebrain. Majority of EAAT4 is expressed in dendritic spines that are in close proximity to astrocytes although there is some expression in close proximity to the synapse as well. EAAT5 is exclusively expressed in the retina and has been identified in both cone and rod photoreceptor terminals and in axon terminals of rod bipolar cells. There are three types of vesicular glutamate transporters known as VGLUTs 1–3. These transporters pack glutamate into synaptic vesicles, so that it can be released into the synapse.

Glutamate Receptor Genes

Due to the diversity of glutamate receptors, their subunits are encoded by numerous gene families. Four GRIA genes code for AMPA subunits, seven GRIN genes code for NMDA subunits and five GRIK genes code for kainate subunits. Finally, metabotropic receptors, which are G-protein coupled receptors, are coded by eight GRM genes. Mutations in genes coding for ionotropic glutamate receptor subunits, which are central to synaptic transmission and plasticity, are mostly associated with intellectual disability and autism spectrum disorders while mutations of metabotropic GluRs, having a role on modulating neural transmission, are preferentially associated with psychiatric disorders.

Glutamate Role in Neuropsychiatric Disorders

Glutamate Role in Aggression

Numerous preclinical studies suggest that stimulation of central glutamate receptors typically increases aggressive behavior in lower mammals, due to the fact that the NMDARs have a central function in the regulation of information processing (10), and aggressive and defensive responses [11]. Preclinical studies of the other ionotropic glutamate receptor, AMPA, and metabotropic glutamate receptors also support a glutamate hypothesis of aggressive behavior. For example, mice deficient for the AMPA receptor GluR-A1 subunit are less aggressive than their wild-type counterparts [12]. In addition, Vekovischeva noted that treatment with AMPA receptor antagonists reduces aggression in aggressive mice strains [13]. Furthermore knockout of the GluA3-AMPA receptor subunit in mice (14), and selective mGlu-1 [15] and mGlu-5 receptor blockade also reduces aggression in mice models of aggression [16]. In contrast, agonist stimulation of auto-inhibitory mGlu-2/3 (17) or mGlu-7 (16) receptors induces the same effect on aggression in mice. As for related clinical findings,

Coccaro et al. found statistically significant direct correlations between CSF glutamate levels and composite measures of aggression, impulsivity, and impulsive aggression in all subjects with personality disorders within their study group and provided initial support for a glutamatergic hypothesis of aggression in human subjects [18]. Recently, it has also been demonstrated that drugs such as valproate and topiramate which act via NMDA and AMPA receptors, have antiaggressive properties. Valproate was noted to reduce aggression in psychiatric patient's instances with poor impulsivity such as personality disorders, adolescent bipolar disorder with high level of aggression and aggression associated with ADHD. As for topiramate it was proven to be successful in reducing aggression in dementia, Tourette syndrome, subjects with histories of substance abuse and antisocial behavior, and also reduced hyperarousal in veterans with PTSD comorbid with alcohol use. Various reports suggest that another anticonvulsant, lamotrigine is efficacious in decreasing aggression in dementia, epilepsy, PTSD and borderline personality disorder, due to its mechanism of action on attenuation of neuronal excitability on presynaptic sites and on postsynaptic AMPA receptors. Lastly, treatment with memantine, a low-potency noncompetitive NMDA antagonist, resulted in improvement in the agitation/aggression symptom cluster in patients with moderate-to-severe Alzheimer's disease [19].

Glutamate Role in ADHD

Compromised function of prefrontal cortex (PFC) and associated neuropsychological deficits found in ADHD children involve executive operations and working memory deficiencies [20]. PFC sends glutamatergic afferents which modulate the nucleus accumbens (NAc) and ventral tegmental area (VTA) dopaminergic neurons functions [21] and is one of the major source of excitatory amino acid (EAA) input to VTA and NAc [22]. It is suggested that glutamate from PFC enhances dopamine cell activity in the VTA, which in turn synergizes with the psychostimulant-induced increase in DA transmission in NAc to promote psychostimulant-induced behavior [23]. Data on neurochemical abnormalities in ADHD patients showed reduced glutamate in the caudate/putamen and anterior cingulate cortex and the medial PFC of the adults [24] and increased glutamate in the cerebellum. The reduction in glutamate may also be age dependent, as studies in ADHD children have reported increased glutamate, including significant increases in the frontal cortex [25] and in the striatum (26). Further contribution to glutamate hypothesis is corroborated by the fact that proven ADHD medication, methylphenidate reverses the impairments related to attention, caused by the NMDA glutamate antagonist dizocilpine [27]. Moreover, memantine, an uncompetitive NMDA receptor antagonist used for the treatment of Alzheimer's disease, was evaluated in an open-label pilot study for its effectiveness in pediatric patients with ADHD. The higher memantine dose used in this study was associated with a higher rate of completion and larger mean improvement on the clinical assessment scales [28]. Lastly, study by Ludolph et al. draw a conclusion that part of atamoxetine efficacy might be due to its antagonism at NMDA receptors [29].

Glutamate Role in Alzheimer's Dementia

It is well known that glutamate, acting via N-methyl-D Aspartate receptors (NMDAR) plays a major role in cognitive processes mainly due to NMDA-evoked long-term potentiation (LTP), which introduces long-lasting strengthening of synaptic transmission and contributes to brain plasticity in terms of the number and strength of individual synapses, which positively correlates with memory formation [30]. Furthermore, postmortem studies have found reduced NMDA receptor density and gene expression in cortex and hippocampus of AD patients. Many neurochemical studies have also provided evidence of a dysfunction of the glutamatergic system in Alzheimer dementia type, such as reduction of glutamic acid brain content [31], reduced receptor binding [32], decreased cortical and hippocampal glutamate uptake due to glutamatergic synapse loss, and resultant overactivation of NMDA glutamate receptors, excitotoxicity and subsequent neurodegeneration [33]. Recently, Aβ peptides were shown to potentiate glutamate release from normal rodent hippocampus (34). Preclinical in vivo studies found that natural human Aβ dimers are able to induce loss of dendritic spines and excitatory synapses through AMPA and NMDA receptors endocytosis, thus resulting in inhibition of long-term potentiation (LTP) and facilitation of long-term depression (LTD) in normal rodent hippocampus [35]. Synaptic dysfunction in AD begins during the preclinical stage of the disease and also includes accumulation of neurotoxic oligomers of the amyloid-β peptide such as highly aggregatable and neurotoxic Aβ42 peptides. Long term neuronal exposure to these oligomers results in permanent spine loss [36]. Other hypothesized mechanisms of neurodegeneration include changes in tau that causes calcineurin-mediated AMPAR internalization in dendritic spines, and ApoE4 induced impairment in glutamatergic neurotransmission, through reduction of ApoE2, NMDA, and AMPA receptor functions. Emerging evidence suggests that the injurious effects of Aβ in AD may be mediated, at least in part, by excessive activation of extrasynaptic or perisynaptic NMDARs containing predominantly NR2B subunits. This is also supported by the fact that memantine, an NMDAR antagonist approved for the treatment of AD, preferentially inhibits extrasynaptic NMDARs (eNMDARs) [37], and imply that even better synapse protection may be achieved by eNMDAR antagonists which are more potent than memantine, and might better protect the brain from glutamate spillover without obstructing normal synaptic transmission [38]. Glutamate uptake/recycling mechanisms are also severely impaired in AD due to deficits in glutamate transporter expression. Glutamate transporter capacity and protein expression is reduced in the frontal/temporal cortex of AD patients, and there is also a selective loss of the vesicular glutamate transporter affecting reuptake of glutamate into synaptic vesicles [39]. Decreased functionality of both excitatory amino acid transporter 2 (EAAT2) and vesicular transporters, may lead to excess glutamate at the excitatory synapse and excessive "background signaling" of glutamate, which is proposed as one of the causes of AD. Another underlying pathophysiology mechanism is neuorinflammation, caused by age-related amyloid plaques accumulation throughout the brain and locally augmented by selective neurofibrillary tangles deposition in hippocampus and entorhinal cortex [40]. This in turn, leads to loss of NMDAR density and loss of NMDAR expressing cells [41] in these regions, which is followed by secondary neuroinflammation due to neuronal loss which contributes to disease progression. This model is supported by the correlation between regional neuroinflammation, but not plaque load, and cognitive abilities [42], evidences of decreased risk of AD in users of non-steroidal anti-inflammatory agents and presence of neuroinflammation in subjects with mild cognitive impairment who later converted to dementia [43]. Additional evidence is coming from an animal model of neuroinflammation, where memantine also protected cholinergic neurons from inflammatory processes [44]. Recent studies have also emphasized role of astrocytes in maintaining metabolic stability of glutamate–glutamine cycle. Due to structural changes in astrocytes, loss of glutamate synthetase was detected in the perisynaptic regions of the neuropil and in astrocytic endfeet where glutamate transporters are also located [45], which may have an effect on synaptic performance [46]. Lastly, recent reports on interactions between various neurotransmitter systems involved in pathophysiology of AD, such as cholinergic and glutamatergic systems, suggest that some cholinergic receptors such as α7nAChRs, which represent a binding site for Aβ, might play a significant role in Aβ-induced glutamate release [47]. These multiple insights into various mechanisms of neurodegeneration in AD, may lead to formation on new additional strategies for beneficial therapeutic interventions.

Glutamate Role in Bipolar Disorder

Altered glutamatergic neurotransmission has been implicated in the pathophysiology of Bipolar Disorder (BD). Most neurochemical studies in humans have noted increased glutamate levels in the plasma, serum, and cerebrospinal fluid of subjects with BD [48]. For instance, both glutamate and Glx (glutamate +glutamine) levels were found to be higher in different limbic areas in individuals with BD during both the depressed and manic phases of the illness [49], but also during periods of euthymia, suggesting that these dysfunctions might represent a possible trait marker of BD [50]. Most current evidence from postmortem studies suggests that apoptosis and oxidative stress play one of the key roles in the pathophysiology of BD, leading to decreased neuronal/glial cell number. Diverse studies have noted alterations in number, density, and size of neurons and glia, particularly oligodendrocytes in the Anterior Cingulate Cortex (ACC), Prefrontal Cortex (PFC) and subcortical areas [51]. These neuropathological findings may reflect the higher vulnerability of oligodendrocytes to insults such as excess glutamate levels leading to excitotoxicity. Furthermore, MRS studies have also reported glutamatergic abnormalities in individuals with BD. For instance, both glutamate and glutamine levels were found to be higher in different limbic areas in individuals with BD during both depressed and manic phases of the illness [49]. Main finding from genetic studies in BD is related to association of GRIN2B gene, which is encoding NR2B subunit of NMDA receptor polymorphisms, with occurrence of BD [52]. This is also supported by the studies that found reductions of white mater integrity in brain cortical regions within the dorsal network in BD patients with GRIN2B risk allele [53] and recent pharmocogenetic observations during STAR*D study by Fabbri et al. [54] indicate that certain glutamatergic gene variants might also affect early antidepressant treatment response in major depressive disorder. Pharmacological studies reinforce the association between BD and the glutamatergic system by reporting that first line agents to treat BD, such as lithium, valproate, carbamazepine, and lamotrigine, also modulate the glutamatergic system [55]. For instance, chronic treatment with lithium has been shown to upregulate synaptosomal uptake of glutamate in mice [56] thus protecting against glutamate induced excitotoxicity [57]. Lithium was also found to enhance VGLUT1 mRNA expression in neurons of the cerebral cortex [58], while long term administration of valproate led to a dose-dependent increase in hippocampal glutamate uptake capacity, by increasing the levels of the glutamate transporters EAAT1 and EAAT2 in hippocampus [59]. Another anticonvulsant lamotrigine, which is currently approved for relapse prevention in BD, probably has the most direct effects on the glutamate system. There is evidence that it inhibits the release of glutamate in the hippocampus of rats [60], increases AMPA subunit receptor expression [61], and blocks voltage-operated sodium channels, thus inducing neuronal membrane stabilization [62]. Lamotrigine also increases AMPAR activity by upregulating surface expression and phosphorylation of GluR1 and GluR2 in hippocampal neurons [63]. Other novel therapeutics with glutamate properties have being tested in double-blind, placebo-controlled studies, such as ketamine in bipolar depression and the results suggest that it has rapid and sustained antidepressant effects, as well as antisuicidal effects. Its effects were manifested within approximately 40 minutes of a single infusion and persisted on average for 3 days post-infusion [64]. Riluzole another glutamatergic drug, currently approved for the treatment of amyotrophic lateral sclerosis, inhibits glutamate release, increases AMPA trafficking by enhancing membrane insertion of GluR1 and GluR2, and increases glutamate reuptake and neurotrophic factor synthesis [65]. It was proven to be effective as add-on to lithium in patients with bipolar depression [66]. However, negative study with add-on to ketamine was reported as well [67]. More recently an open label pilot study found that memantine had significant antimanic effects [68] which additionally support glutamate involvement in pathophysiology of BD.

Glutamate Role in Epilepsy

Glutamate mediated neurotransmission is important in the pathogenesis of epilepsy. Significant increase in extracellular glutamate lead to increases in neuronal activity [69], characterized by an occurrence of seizures resulting from neuronal hyperactivities [70]. Metabotropic glutamate receptors (mGluRs) play a role in the initiation of epileptic discharge and propagation [71]. In particular, group I mGluRs (mGluR1 and mGluR5) are involved in making abnormal synaptic plasticity, which induce long-lasting depolarization and activate neurons to persistently hyperexcitable state [72]. In human temporal lobe epilepsy (TLE), Blumcke, et al. reported upregulated mGluR1, while Notenboom, et al. showed upregulation of mGluR5 receptors in TLE patients, particularly in hippocampal non-sclerosis groups [73]. In pilocarpine-induced epilepsy animal models, mGluR5 protein expression decreased in the hippocampus, and mGluR-mediated hippocampal long term depression (LTD) was also reduced (74). Another group of metabotropic glutamate receptors, group III mGluRs which are mainly functioning as presynaptic autoreceptors and heteroceptors are also of interest, due to their involvement in short- and long-term brain plasticity [75], including microglial activation induced during seizure activity [76]. Studies have shown that early upregulation of NR2B subunit within these receptors, may be involved in the induction and maintenance of epileptogenesis [77]. Therefore creation of subunit-selective antagonists and allosteric modulators for treating acute but also preventing future seizures precipitated by an episode of seizure activity, might prove to be useful therapeutic approach. Furthermore, a glutamate degrading enzyme, glutamine synthetase (GS) has been shown to be deficient in astrocytes in the epileptogenic hippocampal formation in a subset of patients with temporal lobe epilepsy (TLE) [78]. This GS deficiency leads to increased glutamate levels in astrocytes as well as elevated concentrations of extracellular glutamate. This in turn results in impaired uptake of extracellular glutamate, due to a reduced concentration gradient of the amino acid across the plasma membrane [79], release of glutamate into the extracellular space by astrocytes and subsequent excitotoxicity [80]. Another example is Mesial Temporal Lobe Epilepsy (MTLE), one of the most common forms of drug-resistant, localization-related epilepsies in humans. The seizures in MTLE appear to involve a network of temporal lobe and limbic structures including the hippocampal formation, amygdala, entorhinal cortex, lateral temporal neocortex, medial thalamus, and inferior frontal lobes [81]. Furthermore, one of the hallmarks of MTLE is a well-characterized pathology known as hippocampal sclerosis, which is recognized by a pattern of selective neuronal loss and glial alterations in the hippocampal formation [82]. Studies using in vivo brain microdialysis in patients with MTLE have shown that the sclerotic and epileptogenic hippocampal formation contains five-fold higher concentrations of extracellular glutamate interictally, when compared with the non-sclerotic and nonepileptogenic hippocampal formation [83]. Specifically in patients with hippocampal sclerosis (MTLE), the interictal extracellular glutamate concentration is considerably higher than in patients without this pathology (non-MTLE), despite the 60–80% neuronal loss and doubling of glial density in the sclerotic hippocampal formation [84]. Lastly, analysis of surgically resected brain tissue from patients with MTLE revealed a nearly 40% loss of GS protein and enzyme activity in astrocytes in the sclerotic hippocampal formation. As for involvement of ionotropic receptor family in epilepsy, current opinion is that epileptic discharges are generated by the intrinsic excitability properties of neurons, but are synchronized through recurrent excitatory collateral connections primarily involving AMPA receptors while being prolonged by the activity of NMDA receptors. These studies demonstrate the critical role of AMPA receptors in mediating epileptic synchronization and provide a scientific basis for the use of AMPA receptor antagonists to prevent seizures. This is further supported by the evidence that perampanel which selectively inhibits AMPA receptor-mediated synaptic excitation, without affecting NMDA receptors, is efficacious as adjunctive therapy in the treatment of human partial-onset seizures [85]. These recent findings validate the AMPA receptor as a novel target for epilepsy and encourage further development of potent, selective, orally active non-competitive AMPA receptor antagonists.

Glutamate Role in Migraine

Glutamate is the most important excitatory neurotransmitter, and plays a pivotal role in conveying sensory and nociceptive information in the brain and the spinal cord [87]. Involvement of glutamate and its receptors may alter current approach to migraine and reveal a novel nonvasoactive understanding of its pathogenesis, due to glutamate key roles in trigeminovascular activation, central sensitization, and cortical spreading depression (CSD) [88]. The data from animal experimental cortical spreading depression, c-fos protein expression at the trigeminal nucleus caudalis, plasma protein extravasations at level of dura-mater circulation, electrophysiological studies [89] and human studies such as assessment of glutamate concentration in plasma, platelets and cerebrospinal fluid in migraneurs vs. controls, further support this hypothesis. Genetic studies also investigated the role of glutamate receptors, and found that variants in ionotropic, AMPA GRIA1 and GRIA3 genes were specifically associated with Migraine with Aura suggesting a role of these genes in cortical spreading depression, which is retained to be the path physiological mechanism-underlying aura [90]. Animal experiments with nonselective blockade of AMPA and kainate receptors have shown to result in blockade of trigeminocervical complex cell firing in response to electrical stimulation of dural structures in rats [91] and various compounds with these properties, such as tezampanel and LY466195 have being tested in clinical studies [92]. Also memantine an uncompetitive, NMDA antagonist claimed to preserve the critical balance between blocking excessive activity of NMDA neurotransmission and permitting normal physiological function within the glutamatergic synapse, has shown to have inhibitory effect in cortical spreading depression and preliminary reports support its clinical efficacy in migraine as well [93]. Furthermore, ketamine stopped aura and some neurological symptoms, including headache, in some patients with familial hemiplegic migraine [94]. It is also known that chronic administration of well-established prophylactic medications, such as topiramate with its glutamate modulating properties can also inhibit CSD generation [95] and need of its long term use suggests that a gene regulatory effect might also be involved. Lastly, more selective antagonist such as NR2B-selective antagonists (CP-101,606 and Ro25-6981) also inhibit, in a dose-related manner, KCl-induced spreading depression and might be considered as a new therapeutic strategy for migraine, due to better safety profile compared with nonsubtype-selective drugs [96].

Glutamate Role in PTSD

Neuroimaging studies done recently in patients with PTSD are reporting abnormalities in the prefrontal cortex (PFC), hippocampus and amygdala [97] resulting in poor top-down control from the PFC to the amygdala leading to insufficient extinction of fear-related memories [98] and executive control over fear responses. Furthermore, deficient hippocampal PFC signaling may also underlie contextual memory deficits in PTSD, resulting in poor control of conditioned fear responses. Therefore extinction of learned, consolidated fear memories is important aspect of the treatment process and it was shown that enhancement of NMDA receptor function (e.g., indirectly by the glycine receptor partial agonist D-cycloserine, systemically administered or infused into the amygdala) facilitated extinction learning in animal experiments [99]. Clinically, beneficial effects of D-cycloserine in combination with exposure therapy have been reported in other anxiety disorders such as agoraphobia [99], social anxiety disorder [100] and panic disorder [101], although controlled clinical trials looking specifically at the potentially beneficial effects of D-cycloserine on extinction processes in PTSD are still outstanding. Other compounds specifically those that facilitate AMPA receptor signaling might show promise in the treatment of PTSD. In rats, the AMPA receptor potentiator PEPA and glycine transporter inhibitor NFPS facilitated extinction learning for contextual and cued fear respectively [102]. However the most extensively studied glutamate receptor in relation to fear extinction is the NMDA receptor (NMDAR). Systemic or local administration of NMDAR antagonists into the mPFC produced extinction deficits in animal studies [103]. Due to the potential for major side-effects (e.g., excitotoxicity) a general enhancement of NMDAR transmission is clinically undesirable, and more subtle modulation is required. The repertoire of NMDAR subunits permits the assembly of a range of NMDARs with varying dissociable signaling properties. For instance inhibition of GluN2B-containing NMDARs (via ifenprodil or Ro25-6981) disrupts extinction in mice [104]. The most extensively investigated site of NMDARs is L-glycine binding site on the GluN1 subunit. Systemic administration of ligands on this binding site, such as D-serine or D-cycloserine, augments extinction consolidation [105]. First clinical evidence is also supportive to this concept as soldiers with burns who received ketamine during their operations were less likely to develop PTSD compared to those who did not receive ketamine [106]. Further studies in this area are underway and ketamine showed preliminary efficacy for the rapid treatment of PTSD in recent proof of concept randomized clinical trial.

Glutamate Role in Schizophrenia

The hypofunction of glutamatergic transmission has been implicated in schizophrenia [107]. The first evidence supporting this idea came from a finding that phencyclidine and ketamine, two dissociative anesthetics that induce "schizophrenia-like" symptoms in individuals without past history, turned out to be channel blockers of NMDARs [108].

Furthermore, significantly reduced levels of glutamate in cerebrospinal fluid (CSF) were noted [109], while autopsy studies also revealed reduced expression of NMDAR in schizophrenia patients' brain compared with age-matched controls [110]. Consistently, knock out animal model of NMDAR hypofunction by genetic down-regulation of NMDAR expression, shows traits resembling schizophrenia [111], manifested as an increase in locomotor activity and stereotypic motor behaviors [112]. Genetic association studies were also performed to assess functional polymorphisms that may affect glutamatergic transmission in schizophrenia. One of the best studied genes in this respect is GRIN3A gene, which encodes the NR3A subunit of the NMDA receptor. NR3A acts to suppress NMDA receptor activity and is involved in the development of dendritic spines, increases spine density and initiates synapse maturation and memory consolidation during early postnatal neural development, by modulating NMDA receptor activity. Findings from these studies showed that GRIN3A expression levels in the dorsolateral prefrontal cortex were elevated by approximately 30% in schizophrenia patients relative to controls [113], and that rare single nucleotide polymorphisms (SNPs) have been associated with various types of cognitive and negative symptoms and their response to treatment. Regarding involvement of metabotropic glutamate receptors, most studied are mGlu2/3 receptors. Indeed, one of the mGlu2/3 receptor orthosteric agonists, LY379268 inhibited both ketamine-induced increases in gamma oscillations and reduces spontaneous gamma oscillations in the cortex of rats, which are considered to be main electrophysiological variables targeted in schizophrenia, because of their potential role in information transfer between brain regions [114]. Other compounds being tested are positive allosteric modulators preferentially acting on the mGlu2 receptor over the mGlu3 receptor, which reportedly exert antipsychotic effects in animal models [115]. These compounds do not activate the receptor directly, but act at an allosteric site on the receptor to potentiate glutamate-induced activation and produce their effects only in the presence of an endogenous ligand (glutamate) [116], thereby reducing the risk of side effects and tolerance, related to continuous receptor stimulation with orthosteric agonists. One of the orally active mGlu2 receptor positive allosteric modulators is TASP0443294, which exhibits antipsychotic-like activities in animal models. TASP0443294 inhibited not only dopamine-related, but also NMDA receptor-related abnormal behaviors that are considered to be animal models for the positive symptoms and cognitive dysfunction of schizophrenia, respectively [117]. Recently, the first clinical data with mGlu2 receptor positive allosteric modulators (PAMs) were disclosed, as AZD8529 was tested as monotherapy in a phase 2 schizophrenia trial, and was shown not to be effective, in comparison to active comparator, risperidone [118]. In another exploratory phase 2a study in schizophrenia, the mGlu2 PAM JNJ-40411813/ADX71149 (1-butyl-3-chloro-4-(4-phenyl-1-piperidinyl)-2(1H)-pyridinone) met the primary objectives of safety and tolerability. Moreover, patients treated with antipsychotics who experienced residual negative symptoms were identified as a subgroup of patients who may benefit from add-on treatment with JNJ-40411813 [119]. Several reports suggest that glutamatergic receptors may not be the only or the primary glutamatergic molecules affected in schizophrenia [120]. Recent studies have corroborated the view that post-receptor glutamatergic molecules, as well as perturbation of the dopamine glutamate interplay at multiple levels, may play a pivotal role in schizophrenia pathophysiology. This evidence has prompted the investigation of the involvement of postreceptor molecules—including those belonging to the Post-Synaptic-Density (PSD) proteins —in the transduction of glutamatergic signaling and in the cross-talk between other neurotransmitter systems. The

PSD has been defined as an electron-dense thickening at the dendritic spines of glutamatergic postsynapses, a site where glutamate receptors and their signaling downstream effectors are highly concentrated. Whereas NMDA receptors are located within the PSD core, non-NMDA ionotropic and metabotropic glutamate receptors are located at the edge of this ultrastructure [121]. Thus, the disruption of post-receptor clusters and the loss of functional membrane glutamate receptors due to PSD dysfunction may be the cause of complex behavioral deficits (e.g., motor, cognitive deficits). For instance, it has been proposed that scaffolding PSD proteins may subserve synaptic plasticity by regulating trafficking and localization of glutamatergic and non-glutamatergic receptors within the PSD [122] and that disruption of these scaffolds may cause synaptic plasticity dysfunctions leading to schizophrenia-related symptoms. In the PSD, one of the scaffolding proteins are Shank proteins, which bridge glutamate receptors with other scaffolding proteins, cytoskeleton factors and intracellular effectors. Converging reports are highlighting a major role for Shank protein alterations in conditions with a strong genetic component and cognitive impairment, including schizophrenia and schizophrenia-related disorders. Homer proteins, another family of scaffold proteins at the PSD, anchor type I mGluRs to NMDA receptors and bridge mGluRs to their intracellular downstream effectors [123]. Within the PSD, Homer proteins participate in a variety of biological functions, including glutamatergic signaling, regulation of intracellular Ca++signals, cross-talk between different neurotransmitter signaling pathways, and dendritic spine plasticity [124]. Furthermore, a number of single nucleotide polymorphisms (SNPs) within the Homer1gene sequence have been associated with increased scores on the Positive and Negative Syndrome Scale (PANSS) subscales, including total, negative, and positive subscales. These SNPs have also been associated with improvements in PANSS subscales, after administration of antipsychotics [125]. In agreement with these human genetic studies, recent reports have shown that Homer1a expression is consistently altered by antipsychotic treatments [126], suggesting that the modulation of the different Homer transcripts may be relevant to schizophrenia therapy. Accordingly, Homer1 knock-out has been described to impair Long Term Potentiation (LTP) and spatial learning, and exhibit disrupted prepulse inhibition, impaired working memory performance, and increases locomotor response to MK-801 and amphetamine in mice [127]. Another family of proteins are kalirin proteins which are highly enriched in the central nervous system (CNS) and virtually absent elsewhere. The most abundant member of this family is kalirin-7, which is prominently localized in the PSD of excitatory dendritic spines on cortical pyramidal neurons [128]. Kalirin-7 is also thought to regulate the activity-dependent structural remodeling and functional plastic changes within dendritic spines. Moreover, kalirin-7 interacts with several other targets within the PSD and controls a number of postsynaptic functions, including receptor trafficking and vesicle secretion. In cultured neurons, knockdown of kalirin-7 expression reduces the spine content of GluA1 subunits and diminishes the frequency and the amplitude of AMPA receptor-mediated postsynaptic potentials [129]. Concerning the functional significance of kalirin-7 on behavioral and cognitive responses, studies have indeed shown that in kalirin-7 transgenic mice working memory and learning in passive avoidance tasks, are impaired. Kalirin-7 transgenic mice also exhibits reduced prepulse inhibition, diminished social interactions, and increased locomotor activity when compared with wild type mates [130]. Recent research has pointed out that clozapine positive effects on cognitive and negative symptoms might be due to its boosts of neuronal activity in the prefrontal cortex through enhancement of functional synergism between 5-HT1, NMDA receptor, and the Ca2+/calmodulin-dependent protein kinase II (CaMKII), presumably via a CaMKII-mediated recruitment of AMPA receptors to the PSD and via an increase in excitatory postsynaptic potentials [131]. Another potentially useful group of agents might be Phosphodiesterase 10A (PDE10A) inhibitors, proven to be efficacious in preclinical models against positive, negative, and cognitive symptoms, probably by increasing phosphorylation of the GluA1 subunit of AMPA receptors in striatal neurons and at the same time potentiating dopamine D_1 and inhibiting dopamine D_2 receptor signaling [132], thus suggesting that they may represent putative antipsychotic agents targeting postsynaptic sites at both glutamatergic and the dopaminergic systems.

Glutamate Role in Obsessive Compulsive Disorder (OCD)

Glutamate is the primary neurotransmitter within cortico-striatalthalamic circuits (CTSC) with the majority of axon terminals in the striatum being glutamatergic [133]. Increasing evidence has shown that the neurotransmission of glutamate within CTSC circuits is disrupted in OCD [134]. Proton magnetic resonance spectroscopy have revealed greater left but not right caudate glutamatergic (Glx) concentrations [135] in pediatric and adult OCD patients regardless of medication status [136]. Greater orbital frontal Glx has also been reported in adults with OCD [137]. Reduced anterior cingulate cortex (ACC) Glx and increased caudate Glx in pediatric OCD patients may be consistent with a prior report of inverse correlations in OCD patients between ACC and striatal volume [138]. Taken together, these findings suggest a possible tonic-phasic dysregulation of Glx within CTSC circuitry in OCD with reduced tonic Glx in ACC combined with phasic overactivity in the striatum and orbitofrontal cortex [139]. Further evidence for glutamatergic dysfunction in OCD comes from two studies measuring levels of glutamate (and other amino acids) in cerebrospinal fluid (CSF) of OCD patients and matched psychiatrically normal controls [140]. Glutamate concentrations in the CSF of OCD patients were significantly higher compared to normal controls. In animal models, mice with DLGAP3 deletions, displayed OCD like behaviour consisting of compulsive grooming behaviour leading to facial hair loss and skin lesions. DLGAP3 knockout mice were found to have changes in NMDA receptor composition including increased NR1 (GRIN1), NR2B (GRIN2B) subunits and decreased NR2A (GRIN2A) subunits in the postsynaptic density (PSD) of striatal neurons, as well as reduced cortical-striatal transmission. These animal behaviours have good face validity for human grooming disorders, such as trichotillomania (compulsive hair pulling), which frequently co-occur with OCD and are considered by many to form part of an OCD "spectrum" disorders [141]. In these type of animal experiments lever pressing is postulated as a model of the excessive and unreasonable behaviour that characterizes human OCD. The compulsive lever pressing is reduced by SSRIs, but not by drugs known to be less effective for human OCD such as desipramine and diazepam [142]. Interestingly, rats in this model that are treated with D-cycloserine, an NMDA receptor agonist, displayed decreased compulsive lever pressing, whereas the treatment with MK 801, an NMDA antagonist, had no effect on lever pressing [143]. Other abnormality noted in OCD is a genetic coding mutation in exon 12 of SLC1A1, present in close proximity to common variants previously identified in earlier studies of OCD [144]. As for GRIN2B, in a preliminary study, Arnold et al. [145] found a significant association between OCD and variants within a specific 3′ untranslated region in GRIN2B, as well as an even stronger association with a haplotype block in the same region. An independent group has also reported association between GRIN2B variation and OCD, although the specific polymorphisms and haplotypes differed [146].

In a study of glutamatergic concentration (measured using 1H MRS) in children with OCD, Arnold et al. [147] found a significant association between a GRIN2B single nucleotide polymorphism and glutamatergic concentration in the ACC but not the occipital cortex. Another receptors that might be involved in the pathophysiology of OCD are kainate receptors. GRIK2 and GRIK3 genes encode for subunit 2 and 3 of kainite receptors and have been also implicated in genetic association studies of autism and schizophrenia [148]. Delorme et al. [149] reported that GRIK2 SNP I867 (rs2227283) was undertransmitted in OCD trios, supporting a functional role for this variant. To follow up on the finding on GRIK2, Sampaio et al. [150] studied 47 OCD probands and their parents in a family-based genetic association study. While they did not replicate the findings from the Delorme study, they found a different polymorphism, rs1556995 and a two marker haplotype rs1556995/rs1417182 significantly associated with OCD. All this evidence on glutamatergic dysfunction involvement in the pathophysiology of OCD, raised investigators interest in the use of glutamate targeting agents for the treatment of OCD (151). One of them is riluzole which is approved by the FDA for the treatment of Amyotrophic Lateral Sclerosis (ALS). It is thought to reduce glutamatergic neurotransmission through the inhibition of voltage dependent sodium channels [152] and the inhibition of P/Q-type calcium channels at the axon terminal, thereby inhibiting synaptic glutamate release [153] and stimulating glutamate uptake by astrocytes [154]. Several open-label trials have been conducted on riluzole augmentation therapy for both adult and paediatric OCD patients including SSRI-refractory patients and most of them found riluzole to be effective in reducing OCD symptoms (155). Memantine a non-competitive NMDA receptor antagonist approved for the treatment of Alzheimer's disease, has shown efficacy as an augmenting agent in refractory OCD in several case reports [156], as well as open-label trials [157]. Recently, a single blind, case-control study reported memantine was an effective augmenting agent to standard Intensive Residential Treatment (IRT) in severe OCD [158]. Another agent is D-cycloserine (DCS) which has shown to accelerate extinction learning in rodents [159], suggesting it may help facilitate behavioural therapy for OCD where extinction is an important component for reducing anxiety. Two groups have reported successful use of DCS in augmentation with exposure therapy for adult patients [160]. Another recent report showed trend towards improved treatment effect, when DCS was augmented with cognitive behavioural therapy (CBT) in children [161]. Ketamine has also engendered considerable interest recently as a putative agent for OCD treatment with initial case report showing rapid resolution of OCD symptoms [162], while glycine another potentially useful glutamatergic agent has been tested in a small, placebo-controlled study. The results of this preliminary study in which patients were given high doses of glycine orally, previously shown to affect brain glycine levels and NMDA activation, were encouraging [163]. An alternative pharmacological strategy to oral intake of glycine is to enhance glycine tone by reducing its reuptake and/or breakdown in the brain, as glycine is removed from the extracellular space by the glycine transporter Gly-T, primarily by Gly-T1. Gly-T1 inhibitors have received increasing attention recently, primarily in the context of studies of schizophrenia [164]. Sarcosine, or N-methylglycine, a naturally occurring Gly-T1 inhibitor has been explored in this context. A recent open-label study, by Wu and colleagues [165] investigated the potential benefit of sarcosine, either as monotherapy or augmentation, in OCD. This study showed that fast therapeutic effect could be achieved in some OCD patients, particularly those who are treatment naïve. An exciting new possibility is raised by a recent study revealing unexpected glutamate modulating properties of beta lactam antibiotics. Rothstein and colleagues [166] tested over 1000 drugs in an in vitro assay and found that multiple beta-lactam compounds specifically upregulated the glial glutamate uptake transporter. Because of the extensive tolerability of such compounds, they might represent a novel and unexpected group of potential antiglutamatergic agents for use in OCD.

Glutamate Role in Addiction

Addiction is associated with decreased basal prefrontal cortex (PFC) glutamatergic neuronal activity, or "hypofrontality" [167]. Anatomically, the dorsal mPFC is thought to predominantly innervate the nucleus accumbens (NAc) core, while the ventral mPFC sends projections to the NAc shell [168]. The dorsal and ventral mPFC are also thought to play different roles in addiction. Activity in the dorsal mPFC is thought to initiate drug seeking and is crucial for cue-induced relapse to drug seeking [169]. On the other hand, activity in the ventral mPFC can suppress drug-seeking responses [170]. As a whole, the mPFC therefore plays an important role in the control of relapse behavior. Initially, exposure to an addictive drug induces synaptic changes in the ventral tegmental area (VTA). In VTA dopaminergic neurons, drug exposure results in synaptic insertion of high conductance GluA2-lacking AMPARs in exchange for lower conductance GluA2-containing AMPARs. This drug-induced synaptic potentiation in the VTA subsequently triggers synaptic changes in downstream areas of the mesocorticolimbic system (such as the NAc, the PFC, and the amygdala) with further drug exposure. The resulting gradual potentiation of AMPAR transmission and reduced basal extracellular glutamate levels in the NAc mediate incubation of cue-induced craving, by enhancing the responsiveness of NAc medium spiny neurons (MSNs) to drug-associated cues. More specifically in cocaine addiction also a switch of NMDAR subunit composition occurs, due to drug induces creation of NMDARs that have very low Ca2C permeability and are tri-heteromeric GluN1/GluN2B/GluN3A-containing NMDARs instead of previous GluN1/GluN2As and GluN1/GluN2A/ GluN2Bs NMDARs [171]. On the other hand, cocaine drives previously mentioned synaptic insertion of GluA2-lacking AMPARs, which are Ca2C-permeable. As a result of these changes, a form of long term potentiation (LTP) can be induced in VTA neurons that relies on calcium entry through AMPARs and is independent of NMDARs [172]. Ungless et al. [173] demonstrated that AMPAR transmission was enhanced at glutamatergic synapses onto dopaminergic (DA) neurons of the VTA, since both the amplitude and frequency of AMPAR-mediated miniature Excitatory Post Synaptic Currents (mEPSCs) were significantly increased after in vivo exposure to cocaine caused by the insertion of GluA2-lacking AMPARs into the synapse. This gradual potentiation of AMPAR transmission during prolonged withdrawal may mediate "incubation of drug craving," the phenomenon expressed by progressive increase in cue-induced craving and drug seeking over the first months after withdrawal from abused drugs [174]. Dong and Nestler [175] also proposed a hypothesis of rejuvenation of synaptic transmission during cocaine addiction, which manifests through creation of synapses types that are normally associated with early brain development. Examples are the generation of silent synapses and the synaptic insertion of GluA2-lacking AMPARs. This is thought to explain how drugs can elicit such strong and unusually long-lasting plastic changes, leading to equally strong and durable drug-associated memories. The rejuvenation hypothesis also extends to synaptic changes in the VTA, such as the re-emergence of GluN3A-containing NMDARs. Glutamatergic synaptic plasticity does not only occur during drug exposure and withdrawal. It also occurs during cue-induced relapse to drug seeking. The presentation of cocaine-associated cues in the absence of cocaine was found to elicit rapid increases in both the dendritic spine size and the AMPAR/NMDAR ratio in the NAc core of rats that had a history of cocaine self-administration [176]. Furthermore activity in the prelimbic cortex (PL), a subregion of the medial PFC (mPFC) that sends glutamatergic projections to the NAc core, was shown also to be crucially involved in relapse triggering process. It is thought that the impact of cue-induced activity at glutamatergic PL-to NAc synapses is enhanced by the reduction in basal extracellular glutamate levels in the NAc associated with withdrawal from addictive drugs [177]. Moreover basal extracellular glutamate levels in the NAc are also reduced due to downregulated activity of the glial cystine–glutamate exchanger, which transports glutamate to the extracellular space [178]. This results in decreased glutamate tone on presynaptic inhibitory mGluR2 and mGluR3 receptors, which in turn decreases inhibition of presynaptic glutamate release [179] and leads to an increased synaptic release of glutamate, in response to drug associated cues. Furthermore, the released glutamate is not efficiently removed from the synapse, because the expression of the glial glutamate transporter 1 (GLT-1), which mediates glutamate uptake, is diminished following exposure to drugs [180]. All these changes give rise to excessive amounts of glutamate in NAc glutamatergic synapses, which may explain why drug-associated cues have such a strong behavioral effect. In addition to directly modifying synaptic strength in the NAc, drugs of abuse are also thought to alter the capacity of NAc synapses to undergo subsequent normal synaptic plasticity impairments which may contribute to inflexible, compulsive behaviors characteristic of addiction and might explain drug consumption despite negative health consequences in humans. As previously mentioned drug-induced synaptic plasticity in the VTA represents a first step in a cascade of synaptic changes throughout mesocorticolimbic system, therefore an early intervention at the level of the VTA might prevent synaptic changes in downstream brain regions such as the NAc. For instance in rodents, cocaine-induced synaptic plasticity in the VTA can be reversed by activation of mGluR1 receptors [181] that produces a form of mGluR-dependent LTD and normalizes synaptic transmission. Furthermore, this type of mGluR-dependent LTD also reverses the cocaine-induced redistribution of NMDARs, suggesting that mGluR1s might be a potential target to treat early stages of addiction. A different strategy that might prove beneficial in the treatment of addiction would be to rescue drug-induced deficits in NMDAR dependent synaptic plasticity in the NAc, which contribute to inflexible, compulsive behaviors characteristic of addiction. Since the deficits in NMDAR-dependent synaptic plasticity in the NAc are thought to be at least partially due to reduced D-serine levels, enhancing of D-serine signaling by administration of an inhibitor of the enzyme D-amino acid oxidase, may hold promise for the treatment of drug addiction [182]. Given the role of the mPFC in control of relapse behavior, novel treatment strategies could be aimed also at reducing activity in the dorsal mPFC or at enhancing glutamatergic transmission in the ventral mPFC. In particular, agents that inhibit cue-induced endocytosis of GluA2-containing AMPARs might be useful to prevent relapse. Another approach for preventing cue-induced relapse could be to reduce the impact of this activity at glutamatergic mPFC-to-NAc synapses in the NAc by restoring GLT-1 and extracellular glutamate levels. One of the compounds that restores levels of GLT-1 and extracellular glutamate in the NAc is the antioxidant N-acetylcysteine (NAC). NAC treatment has already been tested in human addicts, and was found to reduce drug craving and cocaine use [183]. Another agent that might be useful in the treatment of cocaine addiction is

AMN082 which is positive allosteric modulator (PAM) of mGlu7 receptors [184]. It was shown that it inhibits the primary reinforcing effects of cocaine and thus cocaine consumption, but also its reward enhancing effects, probably by antagonizing rewarding and motivational effects of cocaine on medium spiny GABAergic output neurons in the NAc, which are projecting predominantly to the ventral pallidum [185], and serving as one of the final common pathways mediating psychostimulant reward [186]. Another major challenge in the treatment of drug addiction in humans is the high rate of relapse to drug-taking and drug-seeking behavior even after prolonged periods of abstinence, due to drug craving and reinstatement behaviour [187]. Accordingly, reinstatement of drug-seeking behavior in animals is considered a putative model of relapse in humans [188].

Recent reports have shown that AMN082 administration in rats attenuated both cocaine priming-induced increases in extracellular glutamate in the NAc and reinstatement of cocaine-seeking (189). Another potential approach in relapse prevention would be to target excessive AMPA receptor signaling as administration of AMPA receptor antagonists directly into the accumbens has shown to inhibit reinstatement of ethanol, cocaine and heroin seeking [190]. Several clinical trials with new generation AMPA antagonists are ongoing to evaluate their efficacy in various neurological disorders, although none is being clinically tested in addiction. Previously topiramate an antiepileptic drug with AMPA antagonist and GABA-enhancing effects, has shown positive results in the treatment of alcoholism. Recently also tezampanel (AMPA/kainate receptor antagonist) reduced cocaine self-administration in preclinical studies [191]. Other potentially interesting compounds are talampanel (AMPA/kainate receptor antagonist) and perampanel (noncompetitive AMPA antagonist). There is also extensive evidence that implicates NMDA receptor in extinction of addictive behavioural patterns [192]. For example, enhancement of NMDA receptor function by the NMDA receptor partial agonist D-cycloserine (DCS) facilitates extinction of cocaine CPP and naloxone-induced conditioned place aversion in morphine-dependent rats, and retards subsequent reconditioning to ethanol and cocaine. Two clinical studies have evaluated the effects of DCS in extinction. One study in smokers undergoing extinction exposure showed that DCS facilitated extinction in contrast to subjects receiving placebo [193]. The other study conducted in cocaine abusers evaluated the effects of DCS on extinction of cocaine cue induced craving and showed a trend toward a decrease in craving [194]. Other approach to increase extinction would be to target NR2B-containing NMDA receptors as repeated drug exposures modify the expression of these subunits in the NAc [195]. Selective blockade of NR2B-containing NMDA receptors with ifenprodil in accumbens supported this hypothesis and inhibited conditioned drug craving and reinstatement of cocaine- and heroin-induced drug seeking in preclinical studies [196]. All these basic and clinical research discoveries are justifying the fact that modulation of glutamate neurotransmission is and continues to be one of the most attractive solutions in both prevention and symptom reduction for all major substance use disorders.

Conclusive Remarks

As a conclusion, I would like to say that overall purpose of this book is to increase knowledge relevant to glutamate, foster scientific research activities, which would eventually translate into discovery of novel medications, and ultimately lead to improvement of quality of life in patients with chronic, psychiatric and neurological disorders, thus enhancing their contribution to our society values.

Last but not least, I would like to thank my whole family especially my parents, my brother and my daughter, my Professors, my "dangerous" friend Ole and all of my colleagues including main contributors of the book who always motivated me to progress in life, encouraged my work and wholeheartedly supported publication of this book.

Yours sincerely,

Dr. Zoran M Pavlovic, MD
Psychiatrist,
Worldwide Clinical Trials Inc.
G5, 3 68159 Mannheim, Germany
Email: zoki67pavlovic@yahoo.com

REFERENCES

[1] Traynelis, S.F., Wollmuth, L.P., McBain, C.J., Menniti, F.S., Vance, K.M., Ogden, K.K., et al., (2010). Glutamate receptor ion channels: structure, regulation and function. *Pharmacol. Rev.* 62, 405–96.

[2] Liu,Y., Wong, T.P., Aarts, M., Rooyakkers, A., Liu, L., Lai, T.W., et al., (2007). NMDA receptor subunits have differential roles in mediating excitotoxic neuronal death both in vitro and in vivo. *J. Neurosci.* 27, 2846–57.

[3] Bartlett, T.E., and Wang, Y.T., (2013). The intersections of NMDAR-dependent synaptic plasticity and cell survival. *Neuropharmacology* 74, 59–68.

[4] Bellone, C., and Nicoll, R.A., (2007). Rapid bidirectional switching of synaptic NMDA receptors. *Neuron* 55, 779–78.

[5] Roberts, A.C., Díez-Garcia, J., Rodriguiz, R.M., López, I.P., Luján, R., Martínez-Turrillas, R., et al., (2009). Downregulation of NR3A-containing NMDARs is required for synapse maturation and memory consolidation. *Neuron* 63, 342–56.

[6] Petralia, R.S., Sans,N., Wang, Y.X., and Wenthold, R.J., (2005). Ontogeny of postsynaptic density proteins at glutamatergic synapses. *Mol. Cell. Neurosci.* 29, 436–52.

[7] Paoletti, P., Bellone, C., and Zhou, Q., (2013). NMDA receptor subunit diversity: Impact on receptor properties, synaptic plasticity and disease. *Nat. Rev. Neurosci.*14, 383–400.

[8] Norris, C.M., Korol, D.L., Foster, T.C., (1996). Increased susceptibility to induction of long-term depression and long-term potentiation reversal during aging. *J. Neurosci.* 16, 5382-92.

[9] Iacovelli, L., Bruno, V., Salvatore, L., Melchiorri, D., Gradini, R., et al., (2002). Native group-III metabotropic glutamate receptors are coupled to the mitogen-activated protein kinase / phosphatidylinositol-3- kinase pathways. *J. Neurochem.* 82(2), 216–23.

[10] Daw, N.W., Stein, P.S., Fox, K., (1993). The role of NMDARs in information processing. *Annu. Rev. Neurosci.* 16, 207–22.
[11] Carobrez, A.P., Teixeira, K.V., Graeff, F.G., (2001). Modulation of defensive behavior by periaqueductal gray NMDA/glycine-B receptor. *Neurosci. Biobehav. Rev.* 25, 697–709.
[12] Vekovischeva, O.Y., Aitta-aho, T., Echenko, O., Kankaanpää, A., Seppälä, T., Honkanen, A., et al., (2004). Reduced aggression in AMPA-type glutamate receptor GluR-A subunit-deficient mice. *Genes, Brain Behav.* 3, 253–65.
[13] Vekovischeva, O.Y., Aitta-aho, T., Verbitskaya, E., Sandnabba, K., Korpi, E.R., (2007). Acute effects of AMPA-type glutamate receptor antagonists on intermale social behavior in two mouse lines bidirectionally selected for offensive aggression. *Pharmacol. Biochem. Behav.* 87, 41–9.
[14] Adamczyk, A., Mejias, R., Takamiya, K., Yocum, J., Krasnova, I.N., Calderon, J., et al., (2012). GluA3-deficiency in mice is associated with increased social and aggressive behavior and elevated dopamine in striatum. *Behav. Brain Res.* 229, 265–72.
[15] Navarro, J.F., de Castro, V., Martin-Lopez, M., (2008). JNJ16259685, a selective mGlu(1) antagonist, suppresses isolation-induced aggression in male mice. *Eur. J. Pharmacol.* 586, 217–20.
[16] Navarro, J.F., de Castro, V., Martín-López, M., (2009). Behavioural profile of selective ligands for mGlu7 and mGlu8 glutamate receptors in agonistic encounters between mice. *Psicothema.* 21, 475–9.
[17] Ago, Y., Araki, R., Yano, K., Kawasaki, T., Chaki, S., Nakazato, A., et al., (2012). The selective metabotropic glutamate 2/3 receptor agonist MGS0028 reverses isolation rearing-induced abnormal behaviors in mice. *J. Pharmacol. Sci.* 118, 295–8.
[18] Coccaro, E.F., Lee, R., and Vezinab, P., (2013). Cerebrospinal fluid glutamate concentration correlates with impulsive aggression in human subjects. *J. Psychiatr. Res.* 47(9), 1247–53.
[19] Comai, S., Tau, M., Pavlovic, Z., Gobbi, G., (2012). The psychopharmacology of aggressive behavior: a translational approach: part 2: clinical studies using atypical antipsychotics, anticonvulsants, and lithium. *J. Clin. Psychopharmacol.* 32(2), 237-60.
[20] Faraone, S.V., Biederman, J., (1998) Neurobiology of attention-deficit hyperactivity disorder. *Biol. Psychiatry* 44, 951–8.
[21] Kalivas, P.W., Sorg, B.A., Hooks, M.S., (1993). The pharmacology and neural circuitry of sensitization to psychostimulants. *Behav. Pharmac.* 4, 315–34.
[22] Sesack, S.R., Pickel, V.M., (1992). Prefrontal cortical efferents in the rat synapse on unlabeled neuronal targets of catecholamine terminals in the nucleus accumbens septi and on dopamine neurons in the ventral tegmental area. *J. Comp. Neurol.* 320, 145–60.
[23] Kalivas, P.W., (2000). A role of glutamate transmission in addiction to psychostimulants. *Addict. Biol.* 5, 325– 9.
[24] Perlov, E., Philipsen, A., Hesslinger, B., Buechert, M., Ahrendts, J., Feige, B. et al., (2007). Reduced cingulate glutamate/glutamine-to-creatine ratios in adult patients with attention deficit / hyperactivity disorder - a magnet resonance spectroscopy study. *J. Psychiatr. Res.* 41, 934–4.
[25] Moore, C.M., Biederman, J., Wozniak, J., Mick, E., Aleardi, M., Wardrop, M. et al., (2006). Differences in brain chemistry in children and adolescents with attention deficit

hyperactivity disorder with and without comorbid bipolar disorder: a proton magnetic resonance spectroscopy study. *Am. J. Psychiatry* 163, 316–8.

[26] Carrey, N.J., MacMaster, F.P., Gaudet, L., Schmidt, M.H., (2007). Striatal creatine and glutamate/ glutamine in attention-deficit/hyperactivity disorder. *J. Child. Adolesc. Psychopharmacol.* 17, 11–7.

[27] Rezvani, A.H., Kholdebarin, E., Cauley, M.M., Levin, E.D., (2009). Attenuation of pharmacologically-induced attention impairment by methylphenidate in rats. *Pharmacol. Biochem. Behav.* 92,141–6.

[28] Findling, R.L., McNamara, N.K., Stansbrey, R.J., Maxhimer, R., Periclou, A., Mann, A. et al., (2007). A pilot evaluation of the safety, tolerability, pharmacokinetics, and effectiveness of memantine in pediatric patients with attention-deficit/hyperactivity disorder combined type. *J. Child. Adolesc. Psychopharmacol.* 17, 19–33.

[29] Ludolph, A.G., Udvardi, P.T., Schaz, U., Henes, C., Adolph, O., Weigt, H.U. et al., (2010). Atomoxetine acts as an NMDA receptor blocker in clinically relevant concentrations *Br. J. Pharmacol.* 160, 283–91.

[30] Collingridge, G.L., Singer, W., (1990). Excitatory amino acid receptors and synaptic plasticity. *Trends Pharmacol. Sci.* 11, 290–6.

[31] Procter, A.W., Stirling, J.M., Stratmann, G.C., Cross A.J., Bowen D.M., (1989). Loss of glycine-dependent radioligand binding to the N-methyl-D-aspartate-phencyclidine receptor complex in patients with Alzheimer's disease. *Neurosci. Lett.* 101, 62–6.

[32] Cross, A.J., Slater, P., Simpson, M., Royston, C., Deakin, J.F., Perry, R.H., Perry, E.K., (1987). Sodium dependent d-[3H]aspartate binding in cerebral cortex in patients with Alzheimer's and Parkinson's diseases. *Neurosci. Lett.*79, 213–7.

[33] Cowburn, R., Hardy, J., Roberts, P., Briggs, R., (1988). Presynaptic and postsynaptic glutamatergic function in Alzheimer's disease. *Neurosci. Lett.* 86,109–13.

[34] Kabogo, D., Rauw, G., Amritraj, A., Baker, G., Kar, S., (2010). [beta]-Amyloid-related peptides potentiate K+-evoked glutamate release from adult rat hippocampal slices. *Neurobiol. Aging.* 31, 1164–72.

[35] Shankar, G.M., Li, S., Mehta, T.H., Garcia-Munoz, A., Shepardson, N.E., Smith, I., et al. (2008). Amyloid-beta protein dimers isolated directly from Alzheimer's brains impair synaptic plasticity and memory. *Nat. Med.* 14, 837–42.

[36] Zempel, H., Luedtke, J., Kumar, Y., Biernat, J., Dawson, H., Mandelkow, E., Mandelkow, EM., (2013). Amyloid-β oligomers induce synaptic damage via Tau-dependent microtubule severing by TTLL6 and spastin. *EMBOJ* 32, 2920–37.

[37] Xia, P., Chen, H.S., Zhang, D., Lipton, S.A., (2010). Memantine preferentially blocks extrasynaptic over synaptic NMDA receptor currents in hippocampal autapses. *J. Neurosci.* 30, 11246–50.

[38] Hardingham, G.E., Bading H., (2010). Synaptic versus extrasynaptic NMDA receptor signaling: implications for neurodegenerative disorders. *Nat. Rev. Neurosci.* 11(10), 682–96.

[39] Kirvell, S.L., Esiri, M., Francis, P.T., (2006). Down-regulation of vesicular glutamate transporters precedes cell loss and pathology in Alzheimer's disease. *J. Neurochem.* 98, 939–50.

[40] Braak, H., Braak, E., (1991). Neuropathological stageing of Alzheimer related changes. *Acta. Neuropathol.* 82(4), 239-59.

[41] Bezzi, P., Domercq, M., Brambilla, L., Galli, R., Schols, D., De Clercq, E. et al., (2001). CXCR4-activated astrocyte glutamate release via TNFalpha: amplification by microglia triggers neurotoxicity. *Nat. Neurosci.* 4, 702-10.

[42] Kreisl, W.C., Lyoo, C.H., McGwier, M., Snow, J., Jenko, K.J., Kimura, N., et al., (2013). In vivo radioligand binding to translocator protein correlates with severity of Alzheimer's disease. *Brain* 136, 2228-38.

[43] Yasuno, F., Kosaka, J., Ota, M., Higuchi, M., Ito, H., Fujimura, Y., et al., (2012). Increased binding of peripheral benzodiazepine receptor in mild cognitive impairment-dementia converters measured by positron emission tomography with [^{11}C]DAA1106. *Psychiatry Res.* 203, 67-74.

[44] Willard, L.B., Hauss-Wegrzyniak, B., Danysz, W., Wenk, G.L., (2000). The cytotoxicity of chronic neuroinflammation upon basal forebrain cholinergic neurons of rats can be attenuated by glutamatergic antagonism or cyclooxygenase-2 inhibition. *Exp. Brain Res.* 134, 58–65.

[45] Robinson, S.R., (2001). Changes in the cellular distribution of glutamine synthetase in Alzheimer's disease. *J. Neurosci. Res.* 66, 972–80.

[46] Verkhratsky, A., Olabarria, M., Noristani, H.N., Yeh, C.Y., Rodriguez, J.J., (2010). Astrocytes in Alzheimer's disease. *Neurotherapeutics* 7, 399–412.

[47] Wang, H.Y., Lee, D.H., D'Andrea, M.R., Peterson, P.A., Shank, R.P., Reitz, A.B., (2000). β-Amyloid1-42 binds to α7 nicotinic acetylcholine receptor with high affinity. Implications for Alzheimer's disease pathology. *J. Biol. Chem.* 275(8), 5626–32.

[48] Zarate, C.A. Jr, Machado-Vieira, R., Henter, I., Ibrahim, L., Diazgranados, N., Salvadore, G., (2010). Glutamatergic modulators: the future of treating mood disorders? *Harv. Rev. Psychiatry,* 18, 293–303.

[49] Yüksel, C., Ongur, D., (2010). Magnetic resonance spectroscopy studies of glutamate-related abnormalities in mood disorders. *Biol. Psychiatry* 68, 785–94.

[50] Foland-Ross, L.C., Thompson, P.M., Sugar, C.A., Madsen, S.K., Shen, J.K., Penfold C. et al., (2011). Investigation of cortical thickness abnormalities in lithium-free adults with bipolar I disorder using cortical pattern matching. *Am. J. Psychiatry* 168, 530–9.

[51] Gigante, A.D., Young, L.T., Yatham, L.N., Andreazza, A.C., Nery F.G., Grinberg, L.T., et al., (2011). Morphometric post-mortem studies in bipolar disorder: possible association with oxidative stress and apoptosis. *Int. J. Neuropsychopharmacol.* 14, 1075–89.

[52] Fallin, M.D., Lasseter, V.K., Avramopoulos, D. Nicodemus, K.K., Wolyniec, P.S., McGrath, J.A., et al., (2005). Bipolar I disorder and schizophrenia: a 440-single-nucleotide polymorphism screen of 64 candidate genes among Ashkenazi Jewish case-parent trios. *Am. J. Hum. Genet.* 77, 6, 918–36.

[53] Carter, C.J., (2007). eIF2B and oligodendrocyte survival: where nature and nurture meet in bipolar disorder and schizophrenia? *Schizophr. Bull.* 33, 61, 343–53.

[54] Fabbri, C., Drago, A., Serretti, A., (2013). Early antidepressant efficacy modulation by glutamatergic gene variants in the STAR*D. *Eur. Neuropsychopharmacol.* 23(7), 612–21.

[55] Yatham, L.N., Kennedy, S.H., Schaffer, A., Parikh, S.V., Beaulieu, S., O'Donovan, C., et al., (2009). Canadian Network for Mood and Anxiety Treatments (CANMAT) and International Society for Bipolar Disorders (ISBD) collaborative update of CANMAT

guidelines for the management of patients with bipolar disorder: update, *Bipolar Disord.* 11, 225–55.

[56] Hokin, L.E., Dixon, J.F., Los, G.V., (1996). A novel action of lithium: stimulation of glutamate release and inositol 1,4,5 trisphosphate accumulation via activation of the N-methyl D-aspartate receptor in monkey and mouse cerebral cortex slices. *Adv. Enzyme Regul.* 36, 229–44.

[57] Hashimoto, R., Hough, C., Nakazawa, T., Yamamoto, T., Chuang, D.M., (2002). Lithium protection against glutamate excitotoxicity in rat cerebral cortical neurons: involvement of NMDA receptor inhibition possibly by decreasing NR2B tyrosine phosphorylation. *J. Neurochem.* 80, 589–97.

[58] Moutsimilli, L., Farley, S., Dumas, S., El Mestikawy, S., Giros, B., Tzavara, E.T., (2005). Selective cortical VGLUT1 increase as a marker for antidepressant activity. *Neuropharmacology* 49, 890–900.

[59] Hassel, B., Iversen, E.G., Gjerstad, L., Tauboll, E., (2001). Up-regulation of hippocampal glutamate transport during chronic treatment with sodium valproate. *J. Neurochem.* 77, 1285–92.

[60] Ahmad, S., Fowler, L.J., Whitton, P.S., (2005). Effects of combined lamotrigine and valproate on basal and stimulated extracellular amino acids and monoamines in the hippocampus of freely moving rats. *Naunyn-Schmiedeberg's Arch. Pharmacol.* 371, 1–8.

[61] Du, J., Suzuki, K., Wie, Y., Wang, Y., Blumenthal, R., Chen, Z. et al., (2007). The anticonvulsants lamotrigine, riluzole, and valproate differentially regulate AMPA receptor membrane localization relationship to clinical effects in mood disorders. *Neuropsychopharmacology* 32, 793–802.

[62] Prica, C., Hascoet, M., Bourin, M., (2008). Antidepressant-like effect of lamotrigine is reversed by veratrine: a possible role of sodium channels in bipolar depression. *Behav. Brain Res.* 191, 49–54.

[63] Bhagwagar, Z., Goodwin, G.M., (2005). Lamotrigine in the treatment of bipolar disorder. *Expert Opin. Pharmacother.* 6, 1401–8.

[64] Zarate, C.A. Jr, Brutsche, N.E., Ibrahim, L., Franco-Chaves, J., Diazgranados, N., Cravchik, A., et al., (2012). Replication of ketamine's antidepressant efficacy in bipolar depression: a randomized controlled add-on trial. *Biol. Psychiatry* 71, 939–46.

[65] Mizuta, I., Ohta, M., Ohta, K., Nishimura, M., Mizuta, E., Kuno, S., (2001). Riluzole stimulates nerve growth factor, brain-derived neurotrophic factor and glial cell line-derived neurotrophic factor synthesis in cultured mouse astrocytes. *Neurosci. Lett.* 310, 117–20.

[66] Zarate, C.A., Quiroz, J.A., Singh, J.B., Denicoff, K.D., De Jesus, G., Luckenbaugh, D.A., et al., (2005). An open-label trial of the glutamate-modulating agent riluzole in combination with lithium for the treatment of bipolar depression. *Biol. Psychiatry* 57, 430–2.

[67] Ibrahim, L., Diazgranados, N., Franco-Chaves, J., Brutsche, N., Henter, I.D., Kronstein, P., et al., (2012). Course of improvement in depressive symptoms to a single intravenous infusion of ketamine vs add-on riluzole: results from a 4-week, double-blind, placebo-controlled study. *Nuropsychopharmacology* 37, 1526–33.

[68] Keck, P.E. Jr, Hsu, K.S., Papadakis, K., Russo, J. Jr., (2009). Memantine efficacy and safety in patients with acute mania associated with bipolar I disorder: a pilot evaluation. *Clin. Neuropharmacol.* 32, 199–204.

[69] Meurs, A., Clinckers, R., Ebinger, G., Michotte, Y., Smolders, I., (2008). Seizure activity and changes in hippocampal extracellular glutamate, GABA, dopamine and serotonin. *Epilepsy. Res*. 78, 50–9.

[70] Fisher, R.S., van Emde Boas, W., Blume, W., Elger, C., Genton, P., Lee, P., Engel, J. Jr, (2005). Epileptic seizures and epilepsy: definitions proposed by the International League Against Epilepsy (ILAE) and the International Bureau for Epilepsy (IBE). *Epilepsia* 46, 470–2.

[71] Ure, J., Baudry, M., Perassolo, M., (2006). Metabotropic glutamate receptors and epilepsy. *J. Neurol. Sci*. 247, 1–9.

[72] Merlin, L.R., (2002). Differential roles for mGluR1 and mGluR5 in the persistent prolongation of epileptiform bursts. *J. Neurophysiol.* 87, 621–5.

[73] Notenboom, R.G., Hampson, D.R., Jansen, G.H., van Rijen, P.C., van Veelen, C.W., et al., (2006). Up-regulation of hippocampal metabotropic glutamate receptor 5 in temporal lobe epilepsy patients. *Brain* 129, 96–107.

[74] Cavarsan, C.F., Tescarollo, F., Tesone-Coelho, C., Morais, R.L., Motta, F.L., et al., (2012). Pilocarpine-induced status epilepticus increases Homer1a and changes mGluR5 expression. *Epilepsy Res.* 101, 253–60.

[75] Scanziani, M., Salin, P.A., Vogt, K.E., Malenka, R.C., Nicoll R.A., (1997). Use dependent increases in glutamate concentration activate presynaptic metabotropic glutamate receptors. *Nature* 385, 630–4.

[76] Drage, M.G., Holmes, G.L., Seyfried, T.N., (2002). Hippocampal neurons and glia in epileptic EL mice. *J. Neurocytol.* 31, 681–92.

[77] Woodhall, G., Evans, D.I., Cunningham, M.O., Jones, R.S.G., (2001). NR2B-containing NMDA autoreceptors at synapses on entorhinal cortical neurons. *J. Neurophysiol*. 86, 1644-51.

[78] Eid, T., Thomas, M.J., Spencer, D.D., Rundén-Pran, E., Lai, J. C., Malthankar, G. V., et al. (2004). Loss of glutamine synthetase in the human epileptogenic hippocampus: possible mechanism for raised extracellular glutamate in mesial temporal lobe epilepsy. *Lancet* 363, 28–37.

[79] Tian, GF, Azmi, H, Takano, T, Xu, Q, Peng, W, Lin, J, Oberheim, N, Lou, N, Wang, X, Zielke, HR, Kang, J, Nedergaard, M., (2005). An astrocytic basis of epilepsy. *Nat Med.* 11(9), 973–81.

[80] Spencer, S.S., (2002). Neural networks in human epilepsy: evidence of and implications for treatment. *Epilepsia* 43(3), 219–27.

[81] Gloor, P., (1991). Mesial temporal sclerosis: Historical background and an overview from a modern perspective. In: Luders, H., editor. *Epilepsy Surgery,* Raven Press, New York, 689-703.

[82] Cavus, I., Pan, J.W., Hetherington, H.P., Abi-Saab, W., Zaveri, H.P., Vives K.P., Krystal J.H., Spencer S.S., Spencer D.D., (2008). Decreased hippocampal volume on MRI is associated with increased extracellular glutamate in epilepsy patients. *Epilepsia* 49(8), 1358–66.

[83] During, M.J., Spencer D.D., (1993). Extracellular hippocampal glutamate and spontaneous seizure in the conscious human brain. *Lancet* 341, 1607–10.

[84] Petroff, O.A., Cavus, I., Kim, J.H., Spencer, D.D., (2004). Interictal extracellular glutamate concentrations are increased in hippocampal sclerosis. *Ann Neurol.* 56, 8-43.

[85] Krauss, G.L., Serratosa, J.M., Villanueva, V. Endziniene, M., Hong, Z., French, J., Yang, H., et al., (2012). Randomized phase III study 306: adjunctive perampanel for refractory partial-onset seizures. *Neurology* 78, 1408–15.

[86] Kerling, F., Kasper, B., (2013). Efficacy of perampanel: a review of clinical trial data. *Acta Neurol. Scand.* 127(Suppl. 197), 25–9.

[87] Andreou, A.P., Goadsby, P.J., (2009). Therapeutic potential of novel glutamate receptor antagonists in migraine. *Expert Opin. Investig. Drugs* 18(6), 789–803.

[88] Ma, Q.P., (2001). Co-localization of 5-HT(1B/1D/1F) receptors and glutamate in trigeminal ganglia in rats. *Neuroreport* 12(8),1589–91.

[89] Goadsby, P.J., Lipton, R.B., Ferrari, M.D., (2002). Migraine—current understanding and treatment. *N. Engl. J. Med.* 346(4),257–70.

[90] Formicola, D., Aloia, A., Sampaolo, S., Farina, O., Diodato, D., Griffiths, L.R. et al., (2010). Common variants in the regulative regions of GRIA1 and GRIA3 receptor genes are associated with migraine susceptibility, *BMC Med. Genet.* 11, 103, 1-12.

[91] Andreou, A.P., Storer, R.J., Holland, P.R., Goadsby, P.J., (2006). CNQX inhibits trigeminovascular neurons in the rat: a microiontophoresis study. *Cephalalgia* 26(11),1383.

[92] Sang, C.N., Ramadan, N.M., Wallihan, R.G., Chappell, A.S., Freitag, F.G., Smith, T.R., et al., (2004). LY293558, a novel AMPA/GluR5 antagonist, is efficacious and well-tolerated in acute migraine. *Cephalalgia* 24(7), 596–602.

[93] Cammarata, D., Krusz, J.C., (2005). Memantine: novel mechanism for migraine and headache prophylaxis. *Headache* 45,820.

[94] Kaube, H., Herzog, J., Kaufer, T., Dichgans, M., Diener, H.C., (2000). Aura in some patients with familial hemiplegic migraine can be stopped by intranasal ketamine. *Neurology* 55, 139 –141.

[95] Ayata, C., Jin, H., Kudo, C., Dalkara, T., Moskowitz, M.A., (2006). Suppression of cortical spreading depression in migraine prophylaxis. *Ann. Neurol.* 59, 652– 61.

[96] Chazot, P.L., (2004). The NMDA receptor NR2B subunit: a valid therapeutic target for multiple CNS pathologies. *Curr. Med. Chem.* 11, 389 –96.

[97] Milad, M.R., Rauch, S.L., (2007). The role of the orbitofrontal cortex in anxiety disorders. *Ann. NY. Acad. Sci.* 1121, 546–61.

[98] Koenigs, M., Grafman, J., (2009). Posttraumatic stress disorder: the role of medial prefrontal cortex and amygdala. *Neuroscientist* 15, 540–8.

[99] Ressler, K.J., Rothbaum, B.O., Tannenbaum, L., Anderson, P., Graap, K., Zimand, E., et al., (2004). Cognitive enhancers as adjuncts to psychotherapy: use of D-cycloserine in phobic individuals to facilitate extinction of fear. *Arch. Gen. Psychiat.* 61,1136–44.

[100] Guastella, A.J., Richardson, R., Lovibond, P.F., Rapee, R.M., Gaston, J.E., Mitchell, P., et al., (2008). A randomized controlled trial of D-cycloserine enhancement of exposure therapy for social anxiety disorder. *Biol. Psychiat.* 63, 544–9.

[101] Otto, M.W., Tolin, D.F., Simon, N.M., Pearlson, G.D, Basden, S., Meunier, S.A., et al., (2010). Efficacy of d-cycloserine for enhancing response to cognitive-behavior therapy for panic disorder. *Biol. Psychiat.* 67, 365–70.

[102] Zushida, K., Sakurai, M., Wada, K., Sekiguchi, M., (2007). Facilitation of extinction learning for contextual fear memory by PEPA: a potentiator of AMPA receptors. *J. Neurosci.* 27, 158–66.

[103] Myers, K.M., Carlezon, W.A., Jr., Davis, M. (2011). Glutamate receptors in extinction and extinction-based therapies for psychiatric illness. *Neuropsychopharmacology* 36, 274–93.

[104] Ogden, K.K., Khatri, A., Traynelis, S.F., Heldt, S.A., (2014). Potentiation of GluN2C/DNMDA receptor subtypes in the amygdala facilitates the retention of fear and extinction learning in mice. *Neuropsychopharmacology* 39, 625–37.

[105] Bukalo, O., Pinard, C., Holmes, A., (2014). Mechanisms to medicines: elucidating neural and molecular substrates of fear extinction to identify novel treatments for anxiety disorders. *Br. J. Pharmacol.* 171, 4690–718.

[106] McGhee, L.L., Maani, C.V., Garza, T.H., Gaylord, K.M., Black, I.H., (2008). The correlation between ketamine and posttraumatic stress disorder in burned service members. *J. Trauma* 64(2 Suppl.), 195–8.

[107] Snyder, M.A., Gao, W.J., (2013). NMDA hypofunction as a convergence point for progression and symptoms of schizophrenia. *Front. Cell. Neurosci.* 7, 31.

[108] Javitt, D.C., Zukin, S.R., (1991). Recent advances in the phencyclidine model of schizophrenia. *Am. J. Psychiatry* 148, 1301–8.

[109] Kim, J.S., Kornhuber, H.H., Brand, U., Menge, H.G., (1981). Effects of chronic amphetamine treatment on the glutamate concentration in cerebrospinal fluid and brain: implications for a theory of schizophrenia. *Neurosci. Lett.* 24(1), 93-6.

[110] Moghaddam, B., Krystal, J.H., (2012). Capturing the angel in "angel dust": twenty years of translational neuroscience studies of NMDA receptor antagonists in animals and humans. *Schizophr. Bull.* 38(5), 942-9.

[111] Coyle, J.T., Basu, A., Benneyworth, M., Balu, D., Konopaske, G., (2012). Glutamatergic synaptic dysregulation in schizophrenia: therapeutic implications. *Handb. Exp. Pharmacol.* (213), 267-95.

[112] Kesner, R.P., Hardy, J.D., Calder, L.D., (1981). Phencyclidine and behavior: Sensory-motor function, activity level, taste aversion and water intake. *Pharmacol. Biochem.* Behav. 15(1), 7-13.

[113] Mueller, H.T., Meador-Woodruff J.H., (2004). NR3A NMDA receptor subunit mRNA expression in schizophrenia, depression and bipolar disorder. *Schizophr. Res.* 71, 361–70.

[114] Hiyoshi, T., Hikichi, H., Karasawa, J., Chaki, S., (2014). Metabotropic glutamate receptors regulate cortical gamma hyperactivities elicited by ketamine in rats. *Neurosci. Lett.* 567, 30-4.

[115] Galici, R., Echemendia, N.G., Rodriguez, A.L., Conn, P.J., (2005). A selective allosteric potentiator of metabotropic glutamate (mGlu) 2 receptors has effects similar to an orthosteric mGlu2/3 receptor agonist in mouse models predictive of antipsychotic activity. *J. Pharmacol. Exp. Ther.* 315, 1181–7.

[116] Schaffhauser, H., Rowe, B.A., Morales, S., Chavez-Noriega, L.E., Yin, R., Jachec, C., et al., (2003). Pharmacological characterization and identification of amino acids involved in the positive modulation of metabotropic glutamate receptor subtype 2. *Mol. Pharmacol.* 64,798-810.

[117] Woolley, M.L., Pemberton, D.J., Bate, S., Corti, C., Jones, DN., (2008). The mGlu2 but not the mGlu3 receptor mediates the actions of the mGluR2/3 agonist, LY379268, in mouse models predictive of antipsychotic activity. *Psychopharmacology* (Berl) 196, 431-40.

[118] Litman, R.E., Smith, M.A., Doherty, J., Cross, A., Raines, S., Zukin, S., (2014). AZD8529, a positive allosteric modulator at the MGLUR2 receptor, does not improve symptoms in schizophrenia: a proof of principle study. Presented at NCDEU an Annual meeting of ASCP, 16–19 June 2014, Florida.

[119] De Boer, P., Sinha, V., Hoeben, E., Anghelescu, I.G., Kezic, I., Daly, E, et al., (2013). Characterization of the clinical effect of a positive allosteric modulator of the metabotropic glutamate receptor 2. *68th Annual Scientific Convention of Society of Biological Psychiatry* 73 (Supplement, Issue 9), A1-A8, 1–344.

[120] Bondi, C., Matthews, M., Moghaddam, B., (2012). Glutamatergic animal models of schizophrenia. Curr. Pharm. Des., 18(12), 1593-604.

[121] Sheng, M., Hoogenraad, C.C., (2007). The postsynaptic architecture of excitatory synapses: a more quantitative view. *Annu. Rev. Biochem.* 76, 823-47.

[122] Iasevoli, F., Tomasetti, C., de Bartolomeis, A., (2013). Scaffolding proteins of the post-synaptic density contribute to synaptic plasticity by regulating receptor localization and distribution: relevance for neuropsychiatric diseases. *Neurochem. Res.,* 38(1), 1-22.

[123] Kammermeier, P.J., (2008). Endogenous homer proteins regulate metabotropic glutamate receptor signaling in neurons. *J. Neurosci.* 28(34), 8560-7.

[124] de Bartolomeis, A., Buonaguro, E.F., Iasevoli, F., (2013). Serotonin-glutamate and serotonin-dopamine reciprocal interactions as putative molecular targets for novel antipsychotic treatments: from receptor heterodimers to postsynaptic scaffolding and effector proteins. *Psychopharmacology* (Berl.) 225(1), 1-19.

[125] Spellmann, I., Rujescu, D., Musil, R., Mayr, A., Giegling, I., Genius, J. et al. (2011) Homer-1 polymorphisms are associated with psychopathology and response to treatment in schizophrenic patients. *J. Psychiatr. Res*., 45(2), 234-41.

[126] Iasevoli, F., Fiore, G., Cicale, M., Muscettola, G., de Bartolomeis, A., (2010). Haloperidol induces higher Homer1a expression than risperidone, olanzapine and sulpiride in striatal sub-regions. *Psychiatry Res*. 177(1-2), 255-60.

[127] Szumlinski, K.K., Lominac, K.D., Kleschen, M.J., Oleson, E.B., Dehoff, M.H., Schwarz, M.K., et al., (2005). Behavioral and neurochemical phenotyping of Homer1 mutant mice: possible relevance to schizophrenia. *Genes Brain Behav*. 4(5), 273-88.

[128] Penzes, P., Johnson, R.C., Alam, M.R., Kambampati, V., Mains, R.E., Eipper, B.A., (2000). An isoform of kalirin, a brain-specific GDP/GTP exchange factor, is enriched in the postsynaptic density fraction. *J. Biol. Chem.* 275(9), 6395-403.

[129] Xie, Z., Srivastava, D.P., Photowala, H., Kai, L., Cahill, M.E., Woolfrey, K.M. et al. (2007). Kalirin-7 controls activity-dependent structural and functional plasticity of dendritic spines. *Neuron* 2007, 56(4), 640-56.

[130] Cahill, M.E., Xie, Z., Day, M., Photowala, H., Barbolina, M.V., Miller, C. A., Weiss, C., Radulovic, J., Sweatt, J.D., Disterhoft, J.F., Surmeier, D.J., Penzes, P., (2009). Kalirin regulates cortical spine morphogenesis and disease-related behavioral phenotypes. *Proc. Natl. Acad. Sci. USA.* 106(31), 13058-63.

[131] Purkayastha, S., Ford, J., Kanjilal, B., Diallo, S., Del Rosario Inigo, J., Neuwirth, L., et al., (2012). Clozapine functions through the prefrontal cortex serotonin 1A receptor to

heighten neuronal activity via calmodulin kinase II-NMDA receptor interactions. *J. Neurochem*. 120(3), 396-407.

[132] Grauer, S. M., Pulito, V.L., Navarra, R.L., Kelly, M.P., Kelley, C., Graf, R., et al., (2009). Phosphodiesterase 10A inhibitor activity in preclinical models of the positive, cognitive, and negative symptoms of schizophrenia. *J. Pharm. Exp. Ther*. 331(2), 574-90.

[133] Parent, A., Hazrati, L.N., (1995). Functional anatomy of the basal ganglia. The cortico-basal ganglia-thalamocortical loop. *Brain Res. Brain Res. Rev*. 20, 91–127.

[134] Rotge, J.Y., Aouizerate, B., Tignol, J., Bioulac, B., Burbaud, P., Guehl, D., (2010). The glutamate-based genetic immune hypothesis in obsessive-compulsive disorder. An integrative approach from genes to symptoms. *Neuroscience* 165, 408–17.

[135] Starck, G., Ljungberg, M., Nilsson, M., Jonsson, L., Lundberg, S., Ivarsson, T., et al. (2008). A (1)H magnetic resonance spectroscopy study in adults with obsessive compulsive disorder: relationship between metabolite concentrations and symptom severity. *J. Neural. Transm*. 115, 1051–62.

[136] Yucel, M., Wood, S.J., Wellard, R.M., Harrison, B.J., Fornito, A., Pujol, J., et al. (2008). Anterior cingulate glutamate glutamine levels predict symptom severity in women with obsessive-compulsive disorder. *Aust. NZ J. Psychiatry* 42, 467–77.

[137] Whiteside, S.P., Port, J.D., Deacon, B.J., Abramowitz, J.S., (2006). A magnetic resonance spectroscopy investigation of obsessive-compulsive disorder and anxiety. *Psychiatry Res*. 146, 137–47.

[138] Rosenberg, D.R., Keshavan, M.S., (1998). Bennett Research Award. Toward a neurodevelopmental model of obsessive--compulsive disorder. *Biol. Psychiatry* 43, 23–40.

[139] Rosenberg, D.R., Mirza, Y., Russell, A., Tang, J., Smith, J.M., Banerjee, S.P., et al., (2004). Reduced anterior cingulate glutamatergic concentrations in childhood OCD and major depression versus healthy controls. *J. Am. Acad. Child Adolesc. Psychiatry* 43, 1146–53.

[140] Bhattacharyya, S., Khanna, S., Chakrabarty, K., Mahadevan, A., Christopher, R., Shankar, SK., (2009). Anti-brain autoantibodies and altered excitatory neurotransmitters in obsessive-compulsive disorder. *Neuropsychopharmacology* 34, 489–96.

[141] Feusner, J.D., Hembacher, E., Phillips, K.A., (2009). The mouse who couldn't stop washing: pathologic grooming in animals and humans. *CNS Spectr*. 14, 503–13.

[142] Joel, D., Ben-Amir, E., Doljansky, J., Flaisher, S., (2004). 'Compulsive' lever-pressing in rats is attenuated by the serotonin re-uptake inhibitors paroxetine and fluvoxamine but not by the tricyclic antidepressant desipramine or the anxiolytic diazepam. *Behav. Pharmacol*. 15, 241–52.

[143] Albelda, N., Bar-On, N., Joel, D., (2010). The role of NMDA receptors in the signal attenuation rat model of obsessive-compulsive disorder. *Psychopharmacology* (Berl). 210,13–24.

[144] Arnold, P.D., Sicard, T., Burroughs, E., Richter, M.A., Kennedy, J.L., (2006). Glutamate transporter gene SLC1A1 associated with obsessive-compulsive disorder. *Arch. Gen. Psychiatry* 63, 769–76.

[145] Arnold, P.D., Rosenberg, D.R., Mundo, E., Tharmalingam, S., Kennedy, J.L., Richter, M.A., (2004). Association of a glutamate (NMDA) subunit receptor gene (GRIN2B)

with obsessive-compulsive disorder: a preliminary study. *Psychopharmacology* (Berl.) 174, 530–8.

[146] Stewart, S.E., Fagerness, J., Moorjani, P., Jenike, E., Beattie, K., Illmann, C., et al., (2008). Family-based association between obsessive-compulsive disorder and glutamate receptor candidate genes. WCPG Annual Meeting.

[147] Arnold, P.D., Macmaster, F.P., Richter, M.A., Hanna, G.L., Sicard, T., Burroughs, E., et al., (2009). Glutamate receptor gene (GRIN2B) associated with reduced anterior cingulate glutamatergic concentration in pediatric obsessive-compulsive disorder. *Psychiatry Res.* 172, 136–9.

[148] Ahmad, Y., Bhatia, M.S., Mediratta, P.K., Sharma, K.K., Negi, H., Chosdol, K., et al., (2009). Association between the ionotropic glutamate receptor kainate3 (GRIK3) Ser310Ala polymorphism and schizophrenia in the Indian population. *World J. Biol. Psychiatry,* 1–4.

[149] Delorme, R., Krebs, M.O., Chabane, N., Roy, I., Millet, B., Mouren-Simeoni, M.C., et al., (2004). Frequency and transmission of glutamate receptors GRIK2 and GRIK3 polymorphisms in patients with obsessive compulsive disorder. *Neuroreport,* 15, 699–702.

[150] Sampaio, A.S., Fagerness, J., Crane, J., Leboyer, M., Delorme, R., Pauls, D.L., et al., (2011). Association Between Polymorphisms in GRIK2 Gene and Obsessive-Compulsive Disorder: A Family-Based Study. *CNS Neurosci. Ther.* 17(3), 141-7.

[151] Pittenger, C., Krystal, J.H., Coric, V., (2006). Glutamate-modulating drugs as novel pharmacotherapeutic agents in the treatment of obsessive-compulsive disorder. *NeuroRx.* 3, 69–81.

[152] Urbani, A., Belluzzi, O., (2000). Riluzole inhibits the persistent sodium current in mammalian CNS neurons. *Eur. J. Neurosci.* 12, 3567–74.

[153] Wang, S.J., Wang, K.Y., Wang, W.C., (2004). Mechanisms underlying the riluzole inhibition of glutamate release from rat cerebral cortex nerve terminals (synaptosomes). *Neuroscience* 125, 191–201.

[154] Frizzo, M.E., Dall'Onder, L.P., Dalcin, K.B., Souza, D.O., (2004). Riluzole enhances glutamate uptake in rat astrocyte cultures. *Cell. Mol. Neurobiol.* 24, 123–8.

[155] Grant, P., Lougee, L., Hirschtritt, M., Swedo, S.E., (2007). An open-label trial of riluzole, a glutamate antagonist, in children with treatment-resistant obsessive-compulsive disorder. *J. Child Adolesc. Psychopharmacol.* 17, 761–7.

[156] Hezel, D.M., Beattie, K., Stewart, S.E., (2009). Memantine as an augmenting agent for severe pediatric OCD. *Am. J. Psychiatry* 166(2), 237.

[157] Feusner, J.D., Kerwin, L., Saxena, S., Bystritsky, A., (2009). Differential Efficacy of Memantine for Obsessive Compulsive Disorder vs. Generalized Anxiety Disorder: An Open-Label Trial. *Psychopharmacol. Bull.* 42, 81–93.

[158] Stewart, S.E, Jenike, E.A, Hezel, D.M, Stack, D.E, Dodman, N.H., Shuster, L., et al., (2010). A single-blinded casecontrol study of memantine in severe obsessive-compulsive disorder. *J. Clin. Psychopharmacol.* 30, 34–9.

[159] Walker, D.L., Ressler, K.J., Lu, K.T., Davis, M., (2002). Facilitation of conditioned fear extinction by systemic administration or intra-amygdala infusions of D-cycloserine as assessed with fear-potentiated startle in rats. *J. Neurosci.* 22, 343–51.

[160] Wilhelm, S., Buhlmann, U., Tolin, D.F., Meunier, S.A., Pearlson, G.D., Reese, H.E, et al., (2008). Augmentation of behavior therapy with D-cycloserine for obsessive-compulsive disorder. *Am. J. Psychiatry*, 165, 335.

[161] Storch, E.A, Murphy, T.K., Goodman, W.K., Geffken, G.R., Lewin, A.B., Henin, A., et al., (2010). A preliminary study of D-cycloserine augmentation of cognitive-behavioral therapy in pediatric obsessive-compulsive disorder. *Biol. Psychiatry* 68, 1073–6.

[162] Rodriguez, C.I., Kegeles, L.S., Flood, P., Simpson, H.B., (2011). Rapid resolution of obsessions after an infusion of intravenous ketamine in a patient with treatment-resistant obsessive-compulsive disorder. *J. Clin. Psychiatry* 72(4), 567–9.

[163] Greenberg, W.M., Benedict, M.M., Doerfer, J., Perrin, M., Panek, L., Cleveland, W.L., et al., (2009). Adjunctive glycine in the treatment of obsessive-compulsive disorder in adults. *J. Psychiatr. Res*. 43(6), 664–70.

[164] Hashimoto, K., (2011). Glycine transporter-1: a new potential therapeutic target for schizophrenia. *Curr. Pharm. Des*. 17(2), 112–20.

[165] Wu, P.L., Tang, H.S., Lane, H.Y., Tsai, C.A., Tsai, G.E., (2011). Sarcosine therapy for obsessive compulsive disorder: a prospective, open-label study. *J. Clin. Psychopharmacol.* 31(3), 369–74.

[166] Rothstein, J.D., Patel, S., Regan, M.R., Haenggeli, C., Huang, Y.H., Bergles, D.E, Jin, L., et al., (2005). Beta-Lactam antibiotics offer neuroprotection by increasing glutamate transporter expression. *Nature* 433, 73–7.

[167] Volkow, N.D., Fowler, J.S., Wang, G.J., (2003). The addicted human brain: insights from imaging studies. *J. Clin. Invest.* 111, 1444–51.

[168] Heidbreder, C.A., Groenewegen, H. J., (2003). The medial prefrontal cortex in the rat: evidence for a dorso-ventral distinction based upon functional and anatomical characteristics. *Neurosci. Biobehav. Rev.* 27, 555–79.

[169] Gipson, C.D., Reissner, K.J., Kupchik, Y.M., Smith, A.C.W., Stankeviciute, N., Hensley-Simon, M.E., et al. (2013). Reinstatement of nicotine seeking is mediated by glutamatergic plasticity. *Proc. Natl. Acad. Sci. USA* 110, 9124–9.

[170] LaLumiere, R. T., Smith, K.C., Kalivas, P.W., (2012). Neural circuit competition in cocaine-seeking: roles of the infralimbic cortex and nucleus accumbens shell. *Eur. J. Neurosci.* 35, 614–22.

[171] Yuan, T., Mameli, M., O' Connor, E. C., Dey, P.N., Verpelli, C., Sala, C., et al. (2013). Expression of cocaine-evoked synaptic plasticity by GluN3A-containing NMDA receptors. *Neuron* 80, 1025–38.

[172] Mameli, M., Halbout, B., Creton, C., Engblom, D., Parkitna, J. R., Spanagel, R., et al., (2009). Cocaine-evoked synaptic plasticity: persistence in the VTA triggers adaptations in the NAc. *Nat. Neurosci.* 12, 1036–41.

[173] Ungless, M.A., Whistler, J.L., Malenka, R.C., Bonci, A. (2001). Single cocaine exposure in vivo induces long-term potentiation in dopamine neurons. *Nature* 411, 583–7.

[174] Grimm, J.W., Hope, B.T., Wise, R.A., Shaham, Y., (2001). Neuroadaptation. Incubation of cocaine craving after withdrawal. *Nature* 412, 141–2.

[175] Dong, Y., Nestler, E. J., (2014). The neural rejuvenation hypothesis of cocaine addiction. *Trends Pharmacol. Sci.* 35, 374–83.

[176] Gipson, C.D., Kupchik, Y.M., Shen, H., Reissner, K. J., Thomas, C. A., Kalivas, P.W., (2013). Relapse induced by cues predicting cocaine depends on rapid, transient synaptic potentiation. *Neuron* 77, 867–72.

[177] Gipson, C.D., Kupchik, Y.M., Kalivas, P.W., (2014). Rapid, transient synaptic plasticity in addiction. *Neuropharmacology* 76, 276–86.

[178] Kalivas, P.W., (2009). The glutamate homeostasis hypothesis of addiction. *Nat. Rev. Neurosci.* 10, 561–72.

[179] Shen, H.W., Scofield, M.D., Boger, H., Hensley, M., Kalivas, P.W., (2014). Synaptic glutamate spillover due to impaired glutamate uptake mediates heroin relapse. *J. Neurosci.* 34, 5649–57.

[180] Kasanetz, F., Deroche-Gamonet, V., Berson, N., Balado, E., Lafourcade, M., Manzoni, O., et al. (2010). Transition to addiction is associated with a persistent impairment in synaptic plasticity. *Science* 328, 1709–12.

[181] Bellone, C., Lüscher, C. (2012). Drug-evoked plasticity: do addictive drugs reopen a critical period of postnatal synaptic development? *Front. Mol. Neurosci.* 5, 75.

[182] D'Ascenzo, M., Podda, M. V., Grassi, C. (2014). The role of D-serine as co-agonist of NMDA receptors in the nucleus accumbens: relevance to cocaine addiction. *Front. Synaptic Neurosci.* 6, 16.

[183] Mardikian, P.N., LaRowe, S.D., Hedden, S., Kalivas, P.W., Malcolm, R.J., (2007). An open-label trial of N-acetylcysteine for the treatment of cocaine dependence: a pilot study. Prog. Neuropsychopharmacol. *Biol. Psychiatry* 31, 389–94.

[184] Mitsukawa, K., Mombereau, C., Lotscher, E., Uzunov, D.P., van der Putten, H., Flor, P.J., et al., (2006) Metabotropic glutamate receptor subtype 7 ablation causes dysregulation of the HPA axis and increases hippocampal BDNF protein levels: implications for stress-related psychiatric disorders. *Neuropsychopharmacology* 31, 1112–22.

[185] Groenewegen, H.J., Wright, C.I., Beijer, A.V., (1996). The nucleus accumbens: gateway for limbic structures to reach the motor system? *Prog. Brain Res.* 107, 485–511.

[186] Xi, Z.X., Gardner, E.L., (2008). Hypothesis-driven medication discovery for the treatment of psychostimulant addiction. *Curr. Drug Abuse Rev.* 1, 303–27.

[187] Shaham, Y., Shalev, U., Lu, L., De Wit, H., Stewart, J., (2003). The reinstatement model of drug relapse: history, methodology and major findings. *Psychopharmacology* (Berl.). 168, 3–20.

[188] Epstein, D.H., Preston, K.L., Stewart, J., Shaham, Y., (2006). Toward a model of drug relapse: an assessment of the validity of the reinstatement procedure. *Psychopharmacology* (Berl.) 189, 1–16.

[189] Li, X., Li, J., Gardner, E.L., Xi, Z.X., (2010). Activation of mGluR7s inhibits cocaine-induced reinstatement of drug-seeking behavior by a nucleus accumbens glutamate-mGluR2/3 mechanism in rats. *J. Neurochem.* 114(5), 1368-80.

[190] LaLumiere, R.T., Kalivas, P.W., (2008). Glutamate release in the nucleus accumbens core is necessary for heroin seeking. *J. Neurosci.* 28, 3170–7.

[191] Di Ciano, P., Everitt, B.J., (2004). Direct interactions between the basolateral amygdala and nucleus accumbens core underlie cocaine-seeking behavior by rats. *J. Neurosci.* 24, 7167–73.

[192] Myers, K.M., Carlezon, W.A. Jr, Davis, M., (2011). Glutamate receptors in extinction and extinction-based therapies for psychiatric illness. *Neuropsychopharmacology* 36, 274–93.

[193] Santa Ana, E.J., Rounsaville, B.J., Frankforter, T.L., Nich, C., Babuscio, T., Poling J, et al., (2009). D-Cycloserine attenuates reactivity to smoking cues in nicotine dependent smokers: a pilot investigation. *Drug Alcohol Depend.* 104, 220–7.

[194] Price, K.L., McRae-Clark, A.L., Saladin, M.E., Maria, M.M., DeSantis, S.M., Back, S.E., et al., (2009). D-cycloserine and cocaine cue reactivity: preliminary findings. *Am. J. Drug Alcohol Abuse* 35, 434–8.

[195] Mao, L.M., Wang, W., Chu, X.P., Zhang, G.C., Liu, X.Y., Yang, Y.J., et al., (2009). Stability of surface NMDA receptors controls synaptic and behavioral adaptations to amphetamine. *Nat. Neurosci.* 12, 602–10.

[196] Ma, Y.Y., Cepeda, C., Cui, C.L., (2009). The role of striatal NMDA receptors in drug addiction. *Int. Rev. Neurobiol.* 89, 131–46.

In: Modulators of Glutamatergic Signaling
Editor: Zoran M. Pavlovic

ISBN: 978-1-63483-493-3

Chapter 1

ROLE OF GLUTAMATE IN AGGRESSION

***Cedo D. Miljevic*[1,2,*], *Ana Munjiza*[1] *and Dusica Lecic Tosevski*[1,2]**
[1]Institute of Mental health, Belgrade, Serbia
[2]School of Medicine, University of Belgrade, Belgrade, Serbia

ABSTRACT

The link between glutamate (the main excitatory neurotransmitter in the mammalian central nervous system) and aggression (any behavior directed toward another individual that is carried out with the intent to cause harm) has recently come in focus due to neurobiological research that was carried out mostly on animal models. When it comes to aggression in the human population, we usually differentiate between physical and verbal aggression; indirect and direct aggression; affective versus instrumental and proactive versus reactive aggression. We have now identified all glutamate receptors: metabotropic and four types of ionotropic ligand-gated ion channels called NMDA receptors, AMPA receptors, kainite and delta receptors, with each one of them having several subforms. Glutamate is present as a neurotransmitter in synapses of neural circuits involved in regulation of emotions and aggression: the prefrontal cortex (PFC), amygdala, hypothalamus, anterior cingulate cortex (ACC), insular cortex, ventral striatum, and other interconnected structures.

Preclinical studies showed that NMDA receptor antagonists modulate pro-aggressive behavior in animals, the same as GluR1-containing AMPA receptors, but we must not forget that all families of glutaminergic receptors play a role in the regulation of aggressive behavior in animal models. Recent studies in human population suggest a potential role of NMDA antagonists in the treatment of aggressive behavior in several neuropsychiatric disorders. Even though pharmacogenetic research concerning glutamate receptors in psychiatric disorders with symptomatic aggression is in its infancy, there is still a lot to be explained in this area and more studies on humans are necessary.

Keywords: glutamate, aggression, psychopharmacology

* Corresponding author: cedo.miljevic@yahoo.com.

INTRODUCTION

Aggression and violence are common in almost every modern society today. Moreover, aggressive behavior was also part of everyday functioning of our ancestors (hunter-gatherers) and it is common in many animal species where it has a role in survival. However, on the other side of its extreme forms, aggression is an unsurpassed human tragedy (wars, mass destructions, terrorism, etc.). This chapter will try to explain the neurobiological link between glutamate and aggression.

1. AGGRESSION: DEFINITION, DIFFERENT TYPES AND IMPACT IN PSYCHIATRIC DISORDERS

Aggression encompasses a range of diverse behavioral patterns and is multidimensional in terms of its origins, motivations, expressions and functions [1]. Definitions of aggression refer to "any form of behavior directed toward the goal of harming or injuring another living being who is motivated to avoid such treatment" [2].

Human aggression is any behavior directed at another individual that is carried out with the *proximate* (immediate) intent to cause harm. In addition, the perpetrator must believe that the behavior will harm the target and that the target is motivated to avoid the behavior [3, 4]. *Violence* is aggression that has extreme harm as its goal (e.g., death). All violence is aggression, but many instances of aggression are not violent. For example, one child pushing another off a tricycle is an act of aggression but it is not an act of violence [5].

Aggression has longitudinal psychological and social effects on both the victims and the aggressors, including internalizing and externalizing symptoms [6], loneliness [7] and even suicidal behavior [8].

In order to define different types of aggression among humans and quantify them, an array of different instruments have been developed. For example, Buss and Perry [9] developed the Aggression Questionnaire (AQ), which uses the factor-analytic methods. In the final version of the Aggression Questionnaire, 4 subscales were proposed, whereby physical and verbal aggression represents the behavioral component, anger represents the affective reaction that accompanies aggression, and hostility represents the cognitive aggression component.

The difference between physical and verbal aggression is obvious; the difference between direct and indirect aggression is more difficult to explain: *indirect aggression* is committed without the presence of the target, such as gossiping or telling lies about someone behind his or her back to get them in trouble [10]. The main dichotomies in different types of aggression are: affective versus instrumental, impulsive versus premeditated and proactive versus reactive [11]. *Hostile, emotional or affective aggression* has historically been described as impulsive, thoughtless (i.e., unplanned), driven by anger, having the ultimate motive of harming the target, and occurring as a reaction to some perceived provocation. It is sometimes called affective, impulsive, or reactive aggression. *Instrumental aggression* may be described as a premeditated means of obtaining some goal other than harming the victim, and being proactive rather than reactive [12]. *Impulsive aggression* is usually described as thoughtless (automatic, fast and without consideration of consequences), reactive and affect-laden. In contrast, *premeditated aggression* is usually defined as thoughtful (deliberative, slow, and instrumental), proactive and lacking affect. Reactive aggression (RA) refers to acts committed in negative affective states such as anger or frustration, or in response to provocation. Alternatively, proactive aggression (PA) refers to acts that are motivated by the desire to reach a specific goal [3, 13, 14].

Intense levels of aggression can occur under the influence of alcohol [15, 16.], in response to the absence of an expected reward [17-19.], in response to brief aggressive interactions [20, 21.], at social instigation [19, 21, 22], when aggression serves as a reinforcement [23], or when there is a genetic predisposition [24].

The experience of violent aggressive behavior, or defeat, results in altered neural activity that is relevant to the pathogenesis of many psychiatric disorders [25-27]. Aggressive and defeat experiences prompt transcriptional and translational changes in the cells of the mesocorticolimbic circuits, and several of these neuroadaptive changes appear to mediate the sensitized response to challenges [28].

1.1. Glutamate: Role, Function

Glutamate is the main excitatory neurotransmitter in the mammalian central nervous system (CNS). Nearly all excitatory neurons in the central nervous system are glutamatergic, and it is estimated that over half of all brain synapses release this agent [29].

Glutamate is a nonessential amino acid that does not cross the blood-brain barrier and must be synthesized in neurons from local precursors. The most prevalent glutamate precursor in synaptic terminals is glutamine. Glutamine is released by glial cells and, once within presynaptic terminals, is metabolized to glutamate by the mitochondrial enzyme glutaminase. Glutamate can also be synthesized by transamination of 2-oxoglutarate, an intermediate of the tricarboxylic acid (TCA) cycle. Hence, some of the glucose metabolized by neurons can also be used for glutamate synthesis.

Following its packaging into synaptic vesicles by Mg^{2+}/ATP-dependent transport process, glutamate-occupied vesicles are ready to dock and be released from presynaptic sites. Glutamate is removed from the synaptic cleft by several high-affinity glutamate transporters EAAT, or VGLUT, which are present in both glial cells and presynaptic terminals. In brain injury or disease, transporters can work in reverse, and excess glutamate can accumulate outside cells. This process is called excitotoxicity and it causes calcium ions to enter cells via NMDA receptor channel, which leads to neuronal damage and eventual cell death.

Glial cells contain the enzyme glutamine synthetase, which converts glutamate into glutamine; glutamine is then transported out of the glial cells and into nerve terminals. In this way, synaptic terminals cooperate with glial cells to maintain an adequate supply of the neurotransmitter. This overall sequence of events is referred to as the glutamate-glutamine cycle [30, 31].

1.2. Glutamate Receptors

Glutamate activates metabotropic and several types of ionotropic glutamate receptors. Four of these ionotropic receptors are ligand-gated ion channels called NMDA receptors,

AMPA receptors, kainite and delta receptors. First three glutamate receptors are named after the agonists that activate them: NMDA (N-methyl-D-aspartate), AMPA (α-amino-3-hydroxyl-5-methyl-4-isoxazole-propionate), and kainic acid. The delta receptors, GluD1 and GluD2, are regarded as a subfamily of the ionotropic glutamate receptors solely because of their sequence homology [32].

Glutamate also activates a distinct class of G protein-coupled receptors, called metabotropic glutamate receptors (mGluRs). There are three groups of mGluRs, which are classified as group I (mGluR1, 5), group II (mGluR2, 3), and group III (mGluR4, 6, 7, 8). The group I receptors couple with G protein alpha-q (Gq) to activate the phospholipase C (PLC)-inositol trisphosphate (IP3) cascade. In contrast, group II and III receptors couple with inhibitory Gi/o, which decreases the activity of adenylate cyclase [33].

As a consequence of this diversity of receptor types for glutamate, it is likely that roles of glutamate and its receptor subtypes in aggression differ. Most pharmacological studies of glutamate and aggression primarily focused on the NMDA receptor because of the availability of selective pharmacological agents. However, the role of other receptors in aggression has been gradually discovered by use of gene knockout mice and novel pharmacological tools.

2. Neural Circuits Involved in Aggression and Impact of Glutamate on Them

Neural circuits involved in regulation of emotions including aggression comprise several regions of the prefrontal cortex (PFC), the amygdala, hypothalamus, anterior cingulate cortex (ACC), insular cortex, ventral striatum, and other interconnected structures. Glutamate is present as neurotransmitter in synapses of all these regions.

It have been shown that a linkage of glutamatergic influence in development of aggression with prefrontal cortex (PFC) is associated with dopaminergic transition. Increasing intensity of angry facial expressions is associated with amplified activation of the orbitofrontal cortex (OFC) and the ACC [34]. In two neuroimaging studies that have specifically attempted to induce anger, normal subjects showed augmented activation in the OFC and ACC [35, 36]. Prefrontal cortical glutamate projections excite cells in the ventral tegmental area (VTA) that release dopamine [37-40]. These projections synapse onto dopaminergic cells that ascend back to the prefrontal cortex and onto GABAergic cells that in turn feed back to the nucleus accumbens [41]. This circuit may be essential for regulating the rise in extracellular dopamine in the frontal cortex that occurs during aggressive behavior [42], as well as in other salient experiences [43, 44].

There are two hypothalamic regions of interest with regards to the neural control of attacking in animals: a region situated ventrally and medially to the fornix that induces affective aggression ("defensive rage" in cats, conspecific biting in rats and mice, and outbursts of aggression in humans), and an area lateral to the fornix from where emotionally silent predatory attacks can be induced (attacks on rats in cats and mouse-killing in rats). Hypoarousal associated aggression does not while hyperarousal associated aggression does correlate with increased mediobasal hypothalamic activation [45]. In the hamster anabolic-steroid model, glutamate synthesis are increased in both the anterior and lateral hypothalamus

[46]. It has been speculated that glutamate performs as the aggression output system within lateral hypothalamus [47].

Studies in humans confirm hypothesis of hypothalamic impact in aggression: aggression-related psychopathologies are associated with increased hypothalamic gray matter volume and increased hypothalamic activation during retaliation [48]. Findings suggest that the delivery of severe punishment correlates with increased hypothalamic activation in angry psychopaths but not in angry healthy subjects. Study of Veit et al. shows that hypothalamic activation in psychopaths correlates positively with the amount of punishment delivered in response to the increased punishment received from a yoked opponent; the response also correlated with heightened anger [49]. Other (previous) studies did not confirm these results [50]. Increased gray matter in the hypothalamus of borderline personality patients correlated positively with the history of traumatization [51]. As the aggressiveness of subjects (similar to borderline personality disordered patients in general) showed a more than two fold increase, this finding tentatively suggests that traumatic experience results in an increased hypothalamic gray matter, which may contribute to the expression of aggressiveness. Unfortunately, there are no available studies that correlate glutamate levels in humans with hyperarousal of hypothalamus.

In a number of preclinical studies in rodents, it was demonstrated that an excitatory amino acid pathway from the medial hypothalamus (MH) to the periaqueductal gray matter (PAG) is associated with aggressive behavior [52, 53]. In the rat, there is a dense and distinct group of glutamatergic neurons expressing glutamate transporter protein over the entire hypothalamic attack area, with the rostral part predominantly containing glutamatergic and the caudal part having both glutaminergic and, to a lesser degree, GABAergic, neurons [54].

Excitatory inputs from the basal amygdala project to the PAG and there is evidence for PAG NMDA receptor-mediated defensive rage following their stimulation [55]; basal amygdaloid neurons projecting to PAG defensive-rage sites also stain immunopositive for glutamate. In addition, mice bred for reduction of function in the NMDA R1 subunit display an absence of species-typical fight response in the resident intruder model of aggression [56].

Disruption of the serotonin (5-HT) system has been linked to aggression and violence by a variety of methods [57]. Considerable anatomical, electrophysiological, and neurochemical evidence shows that glutamate interacts with serotonin: in the raphe nuclei via NMDA receptors, via AMPA receptors in the dorsal raphe nucleus (DRN), and via metabotropic receptors in the PAG and the frontal cortex [58-61].

The Medial PFC-DRN glutamatergic projection has been emphasized in pathway-regulating aggression [62]. DRN receives glutamate input by the descending projections from the lateral habenula, periaqueductal gray, lateral hypothalamus, interpeduncular nucleus and from medial prefrontal cortex [63-65]. Both the N-metyl-D-aspartate (NMDA) and α-amino-3- hydroxy-5-methyl-4-isoxazolaproprionate/kainate (AMPA/kainate), ionotropic receptors for glutamate, are localized on serotonergic neurons and increase the 5-HT release in the DRN and its projection areas [66, 67]. Metabotropic receptors of glutamate (mGluRs) are also found in the DRN, but their function on serotonergic neurotransmission and aggressive behavior needs to be characterized by further research.

3. Preclinical Studies Exploring Role of Glutamate in Aggression

3.1. NMDA (*N*-Methyl-D-Aspartate) Receptors

The available preclinical evidence indicates that systemic administration of classic antagonists of NMDA receptors yield highly variable effects in experimental models of aggression [68]. Phencyclidine (PCP) and dizocilpine (MK 801) have been reported to increase levels of aggression. Aggression arises when isolated mice confront an intruder for the first time [69 – 71] or fight him at low levels [72, 73], as well as when rats are sleep deprived [74]. Conversely, in mice and rats with a robust repertoire of aggressive behavior and particularly those that fight at very high, escalated levels, PCP and dizocilpine (MK-801) have been shown to decrease aggression [75-78]. These studies suggest that individuals with very little fighting experience may be more susceptible to the aggression-heightening effects of NMDA receptor antagonists [79].

Memantine and neramaxane block the channel of NMDA receptors with different characteristics and promote alcohol-heightened aggression in resident mice that confront an intruder. Anatomically discrete analysis is required to identify the sites of action for NMDA receptors that produce escalated aggressive behavior [80].

Administration of low-dose methylphenidate (MPH) selectively potentiated NMDA receptor-mediated excitatory postsynaptic currents (EPSCs) via adrenergic receptor activation, whereas high-dose MPH suppressed both NMDAR-mediated and AMPA receptor-mediated EPSCs. The dual effects of MPH on EPSCs were associated with bidirectional changes in the surface level of glutamate receptor subunits; animals injected with high-dose MPH exhibited significantly elevated locomotive activity [81].

Elevated levels of aggressive behavior in animals after administration of NMDA receptor antagonists may be related to their effects on the dopaminergic system. PCP, ketamine, MK-801, memantine have all been shown to increase extracellular concentrations of dopamine in the nucleus accumbens [82-84] and prefrontal cortex [85-87]. Impact of NMDA antagonists on amygdale, medial hypothalamus (MH), periaqueductal gray matter (PAG) pathway can also explain promotion of pro-aggressive behavior in animals after their administration. This glutaminergic pathway has been proposed to mediate the electrical stimulation of "defensive rage" [88, 89]. Pretreatment with the glutamate antagonist – kynurenic acid – blocked medial hypothalamus (MH) facilitation of periaqueductal gray matter (PAG)-elicited defensive rage, and NMDA injected into (PAG) defensive rage sites facilitated the rage response elicited from that site [90]. Administration of an NMDA receptor antagonist into the PAG blocked MH facilitation of PAG-elicited defensive rage. The antagonist suppressed defensive rage elicited by stimulation of the MH, depending on dose [91]. This study also reported that a considerable number of glutamate neurons within the anteromedial hypothalamus project to PAG defensive rage sites.

3.2. AMPA (α-Amino-3-Hydroxyl-5-Methyl-4-Isoxazole-Propionate), Kainite, Metabotropic and Delta Receptors

Fewer studies have been done exploring the link between this group of receptors and aggression, compared to NMDA receptor. Preclinical studies that have been done on AMPA receptors support a glutamate hypothesis of aggressive behavior. GluR1-containing AMPA receptors seem to promote aggressive behavior. It has been shown that mice deficient in the AMPA receptor GluR-A1 subunit are less aggressive than their wild-type counterparts [92] and treatment with AMPA receptor antagonists also reduces aggression in aggressive mice strains [93]. Contrary to this, GluA3-deficiency in mice is associated with increased social and aggressive behavior and elevated dopamine in the striatum [94]. Genome-wide scans to identify quantitative trait loci for aggression in aggressive mice strains find that the Gria3 gene, which encodes for a subunit of the AMPA3 receptor, accounts for the strain differences in aggressive behavior in the resident-intruder mouse model of aggression [95].

When it comes to metabotropic receptors, reduced aggression in mice models was observed after blockade of the receptor for selective mGlu-1 [96] and mGlu-5 [97]. On the other hand, in alcohol-treated animals, a low dose of mGluR5 antagonist MTEP escalated aggressive behavior [98]. Other studies found that agonist stimulation of auto-inhibitory mGlu-2/3 [99] or mGlu-7 [100] receptors reduces aggression in mice. It is possible that group III mGluRs have a similar inhibitory effect on aggression as that of group II mGluRs, because both mGluR groups couple with inhibitory Gi/o. However, their action site varies due to the different localization of these mGluRs [101].

Deletion of the GluR6 gene subunit of kainite receptors in mice induced manic-like behavioral changes, and GluR6 knockout mice are hyperactive, more aggressive toward other males, and exhibit less depressive-like behavior [102].

The role of delta receptors in aggression remains poorly understood. A recent study has shown that mice with deleted GluD1 displayed an increased number of attacks and shorter attack latency compared to characteristically nonaggressive wild-type mice in the resident-intruder test after 3 weeks of isolation. Also, the expression of the GluR1 (AMPA receptor) and GluR6 (kainate receptor) subunits in the amygdale is increased – which promotes aggressive behavior [103]. It is also possible that activation of the GluD1 receptor itself may have a role in regulating aggression circuits [33].

4. Clinical Studies Exploring Role of Glutamate in Aggression

Excessive neuronal discharges (i.e., ictal responses) and the display of episodic violent behavior (i.e., the dyscontrol syndrome) share a common mechanism, presumably involving glutamatergic activity in the temporal lobe [104, 105]. Violent offenders have reduced activity and volume [106, 107] as well as function [108] of the prefrontal cortex. In children with temporal lobe epilepsy, only those with a history of aggression had reduced prefrontal cortex gray matter [109], suggesting that it is prefrontal cortical mediation of limbic structures, rather than excessive temporal lobe activity itself, that contributes to aggressive behavior [79]. One may hypothesize that glutamate may be sensitizing an individual to become more aggressive. Blockage of the glutamate receptors could prevent this sensitized response. In a preliminary study, hospitalized children who were given amantadine, a very low-affinity receptor antagonist that also affects the DA system, to treat impulsivity, aggression, and/or hyperactivity were judged by hospital staff to have fewer of these symptoms [110].

The only study that investigated the relationship between the central nervous system glutamate levels and aggression and/or impulsivity in human subjects showed that glutamate levels did not fluctuate between the personality disorder and healthy subjects but did directly correlate to composite measures of both aggression and impulsivity and a composite measure of impulsive aggression in both groups [111]. However, CSF and brain glutamate levels have been investigated in subjects with neuropsychiatric disorders such as schizophrenia, mood disorder, obsessive compulsive disorder, and pathological gambling [112-118].

Recent studies have highlighted the potential of low-dose or low-potency NMDA antagonists as efficient anti-aggressive treatments in several neuropsychiatric disorders [119, 120].

Conclusion

Links between glutamate and aggression include the same, more unmentioned biological variables like neuroendocrine axis, neuroimunological and genetic aspects. Increased glutamate and vasopressin activity in latero- andanterior hypothalamus facilitate anabolic androgenic steroid (AAS)-induced offensive aggression [122]. Recent studies have suggested an important relationship linking cytokines, immunity and aggressive behavior. Clinical reports describe increasing levels of hostility, anger, and irritability in patients who receive cytokine immunotherapy. Interleukin 1 facilitates defensive rage when microinjected into either the medial hypothalamus or PAG. In contrast, interleukin 2 microinjected into the medial hypothalamus suppresses defensive rage and this suppression is mediated through GABAA receptors. Future researches should explain possible signaling pathways involving glutamate in this process [123], as well as the role of glutamate in neuroimmunology. Recent studies showed that concentrations of glutamate in the brain are linked with individual genetic variations. A pathway approach was established that identified a module composed of 70 genes with high relevance to glutamate biology: individuals carrying a higher number of associated alleles from genes in this module showed the highest levels of glutamate [124]. Association of these genes and aggressive behavior is to be tested. Even though pharmacogenetic research concerning glutamate receptors in psychiatric disorders with symptomatic aggression is in its infancy [125], there is still a lot to be explained in this area and more studies on humans are necessary.

References

[1] Miczek, K. A., Faccidomo, S., Fish, E. W. & De Bold, J. F. (2007). Neurochemistry and Molecular Neurobiology of Aggressive Behavior. In: Blaustein, J., (Ed) *Behavioral*

Neurochemistry, Neuroendocrinology, Molecular Neurobiology. New York: Springer, 285-336.

[2] Baron, R. A. & Richardson, D. R. (1994). Human aggression. New York: Plenum Press,

[3] Bushman, B. J. & Anderson, C. A. (2001). Is it time to pull the plug on the hostile versus instrumental aggression dichotomy? *Psychol. Rev. 108*, 273-79.

[4] Geen, R. G. (2001). Human Aggression. *Taylor & Francis. 2nd ed.*

[5] Anderson, C. A. & Bushman, B. J. (2002). Humman Aggression. *Annu. Rev. Psychol. 53*, 27-51.

[6] Parker, J. G., Rubin, K. H., Erath, S. A., Wojslawowicz, J. C. & Buskirk, A. A. (2006). Peer relationships, child development, and adjustment: A developmental psychopathology perspective. In Cicchetti, D., Cohen, D.J., (Eds.), Developmental psychopathology: Vol. 1. *Theory and method, 2nd ed.*, 419-493. Hoboken.

[7] Schinka, K. C., van Dulmen, M. H., Mata, A. D., Bossarte, R. & Swahn, M. (2013). Psychosocial predictors and outcomes of loneliness trajectories from childhood to early adolescence. *Journal of Adolescence.*, *36*, 1251-1260.

[8] Gvion, Y. & Apter, A. (2011). Aggression, impulsivity, and suicide behavior: A review of the literature. *Archives of Suicide Research.*, *15*, 93-112.

[9] Buss, A. H. & Perry, M. (1992). The aggression questionnaire. *Journal of Personality and Social Psychology.*, *63*, 452-459.

[10] Kirsti, M. J., Björkqvist, L. K. & Peltonen, T. (1988). Is indirect aggression typical of females? gender differences in aggressiveness in 11- to 12-year-old children. *Aggressive behavior.*, *14*, 403-14.

[11] Hogg, M. A. & Cooper, J. (2003). *The Sage Handbook of social psychology*, Chapter 14 Thousand Oaks, CA: Sage Publications, Inc.

[12] Berkowitz, L. (1993). Pain and aggression: some findings and implications. *Motiv. Emot.*, *17*, 277-93.

[13] Poulin, F. & Boivin, M. (2000). Reactive and proactive aggression: Evidence of a two-factor model. *Psychological Assessment.*, *12*, 115-122.

[14] Miller, J. D. & Lynam, D. R. (2006). Reactive and proactive aggression: Similarities and differences. *Personality and Individual Differences.*, *41*, 1469-1480.

[15] Winslow, J. T. & Miczek, K. A. (1985). Social status as determinant of alcohol effects on aggressive behavior in squirrel monkeys (*Saimiri sciureus*). *Psychopharmacology.*, *85*, 167–172.

[16] Miczek, K. A. Barros, H. M., Sakoda, L. & Weerts, E. M. (1998). Alcohol and heightened aggression in individual mice. *Alcohol Clin Exp Res.*, *22*, 1698-1705.

[17] Dollard, J., Doob, L., Miller, N., Mowrer, O. & Sears, R. (1939). Frustration and Aggression. New Haven: Yale University Press

[18] Cherek, D. R. & Heistad, G. T. (1971). Fixed-interval induced aggression. *Psychon Sci.*, *25*, 7-8.

[19] de Almeida, R. M. M. & Miczek, K. A. (2002). Aggression escalated by social instigation or by discontinuation of reinforcement ("frustration") in mice: inhibition by anpirtoline—a 5-HT1B receptor agonist. *Neuropsychopharmacology.*, *27*, 171 181.

[20] Leshner, A. I. & Nock, B. L. (1976). The effects of experience on agonistic responding: an expectancy theory interpretation. *Behav Biol.*, *17*, 561-566.

[21] Potegal, M. & Tenbrink, L. (1984). Behavior of attack-primed and attack-satiated female golden hamsters (*Mesocricetus auratus*). *J Comp Psychol.*, *98*, 66-75.

[22] Fish, E. W., Faccidomo, S. & Miczek, K. A. (1999). Aggression heightened by alcohol or social instigation in mice: reduction by the 5-HT1B receptor agonist CP-94,253. *Psychopharmacology.*, *146*, 391-399.

[23] Fish, E. W., De Bold, J. F. & Miczek, K. A. (2002). Aggressive behavior as a reinforcer in mice: activation by allopregnanolone. *Psychopharmacology.*, *163*(3-4), 459-466.

[24] de Boer, S. F., Van Der Vegt, B. J. & Koolhaas, J. M. (2003). Individual variation in aggression of feral rodent strains: a standard for the genetics of aggression and violence? *Behav Genet.*, *33*(5), 485-501.

[25] Post, R. M. (1992). Transduction of psychosocial stress into the neurobiology of recurrent affective disorder. *Am J Psychiatry.*, *149*, 999-101.

[26] Sinha, R. (2001). How does stress increase risk of drug abuse and relapse? *Psychopharmacology.*, *158*(4), 343-359.

[27] Romans, L., Fresán, A., Senties, H., Sarmiento, E., Berlanga, C., Robles-García, R. & Tovilla-Zarate, C. A. (2014). Validation of the Impulsive / Premeditated Aggression Scale in Mexican psychiatric patients. *Nord J Psychiatry.*, *26*, 1-6.

[28] Miczek, K. A., Covington, H. E., III, Nikulina, E. M. & Hammer, R. P. Jr., (2004). Aggression and defeat: persistent effects on cocaine self-administration and gene expression in peptidergic and aminergic mesocorticolimbic circuits. *Neurosci Biobehav Rev.*, *27*, 787-802.

[29] Purves, D., Augustine, G. J., Fitzpatrick, D., Katz, L. C., LaMantia, A. S., McNamara, J. O. & Williams, S. M., (2001). Chapter 6. Neurotransmitters: Glutamate in Neuroscience, 2nd edition. *Sinauer Associates Sunderland USA*

[30] Meldrum, B. S. (2000). Glutamate as a neurotransmitter in the brain: Review of physiology and pathology. *The Journal of nutrition.*, *130* (4S Suppl), 1007S-1015S.

[31] McEntee, W. J. & Crook, T. H. (1993). Glutamate: Its role in learning, memory, and the aging brain." *Psychopharmacology.*, *111*(4), 391-401.

[32] Orth, A., Tapken, D. & Hollmann, M. (2013). The delta subfamily of glutamate receptors: characterization of receptor chimeras and mutants. *Eur J Neurosci.*, *37*(10), 1620-30.

[33] Kew, J. N. & Kemp, J. A. (2005). Ionotropic and metabotropic glutamate receptor structure and pharmacology. *Psychopharmacology.*, *179*, 4-29.

[34] Blair, R. J., Morris, J. S., Frith, C. D., Perrett, D. I. & Dolan, R. J. (1999). Dissociable neural responses to facial expressions of sadness and anger. *Brain.*, *122*(5), 883-93.

[35] Dougherty, D. D., Shin, L. M., Alpert, N. M., Pitman, R. K., Orr, S. P., Lasko, M., Macklin, M. L., Fischman, A. J. & Rauch, S. L. (1999). Anger in healthy men: a PET study using script-driven imagery. *Biol Psychiatry.*, *46*(4), 466-72.

[36] Kimbrell, T. A., George, M. S., Parekh, P. I., Ketter, T. A., Podell, D. M., Danielson, A. L., Repella, J. D., Benson, B. E., Willis, M. W., Herscovitch, P. & Post, R. M. (1999). Regional brain activity during transient self-induced anxiety and anger in healthy adults. *Biol Psychiatry.*, *46*(4), 454-65.

[37] Sesack, S. R. & Pickel, V. M. (1992). Prefrontal cortical efferents in the rat synapse on unlabeled neuronal targets of catecholamine terminals in the nucleus accumbens septi and on dopamine neurons in the ventral tegmental area. *J Comp Neurol.*, *320*(2), 145-160.

[38] Johnson, S. W., Seutin, V. & North, R. A. (1992). Burst firing in dopamine neurons induced by *N* methyl-Daspartate: role of electrogenic sodium pump. *Science.*, *258*(5082), 665-667.

[39] Tong, Z. Y., Overton, P. G. & Clark, D. (1996). Stimulation of the prefrontal cortex in the rat induces patterns of activity in midbrain dopaminergic neurons which resemble natural burst events. *Synapse.*, *22*(3), 195-208.

[40] Takahata, R. & Moghaddam, B. (2000). Target-specific glutamatergic regulation of dopamine neurons in the ventral tegmental area. *J Neurochem.*, *75*(4), 1775-1778.

[41] Carr, D. B. & Sesack, S. R. (2000). Projections from the rat prefrontal cortex to the ventral tegmental area: target specificity in the synaptic associations with mesoaccumbens and mesocortical neurons. *J Neurosci.*, *20*(10), 3864-3873.

[42] Van Erp, A. M. M. & Miczek, K. A. (2000). Aggressive behavior, increased accumbal dopamine and decreased cortical serotonin in rats. *J Neurosci.*, *15*, 9320-9325.

[43] Pfaus, J. G., Damsma, G., Nomikos, G. G. et al., (1990). Sexual behavior enhances central dopamine transmission in the male rat. *Brain Res.*, *530*, 345-348.

[44] Bassareo, V. & Di Chiara, G. (1997). Differential influence of associative and nonassociative learning mechanisms on the responsiveness of prefrontal and accumbal dopamine transmission to food stimuli in rats fed *ad libitum*. *J Neurosci.*, *17*(2), 851-861.

[45] Haller, J. (2013). The neurobiology of abnormal manifestations of aggression--a review of hypothalamic mechanisms in cats, rodents, and humans. *Brain Res Bull.*, *93*, 97-109.

[46] Carrillo, M., Ricci, L. A. & Melloni, Jr. R. H. (2009). Adolescent anabolic androgenic steroids reorganize the glutamatergic neural circuitry in the hypothalamus. *Brain Research.*, *1249*, 118-127.

[47] Carrillo, M., Ricci, L. A. & Melloni, R. H. (2011). Glutamate vasopressin interactions and the neurobiology of anabolic steroidinduced offensive aggression. *Neuroscience.*, *185*, 85-96.

[48] Lyons-Ruth, K., Choi-Kain, L., Pechtel, P., Bertha, E. & Gunderson, J. (2011). Perceived parental protection and cortisol responses among young females with borderline personality disorder and controls. *Psychiatry Research.*, *189*, 426-432.

[49] Veit, R., Lotze, M., Sewing, S., Missenhardt, H., Gaber, T. & Birbaumer, N. (2010). Aberrant social and cerebral responding in a competitive reaction time paradigm in criminal psychopaths. *Neuroimage.*, *49*, 3365-3372.

[50] Lotze, M., Veit, R., Anders, S. & Birbaumer, N. (2007). Evidence for a different role of the ventral and dorsal medial prefrontal cortex for social reactive aggression: an interactive fMRI study. *Neuroimage.*, *34*, 470-478.

[51] Kuhlmann, A., Bertsch, K., Schmidinger, I., Thomann, P. A. & Herpertz, S. C. (2013). Morphometric differences in central stress-regulating structures between women with and without borderline personality disorder. *J Psychiatry Neurosci*, *38*(2), 129-37.

[52] Beitz, A. J. (1989). Possible origin of glutamatergic projections to the midbrain periaqueductal gray and deep layer of the superior colliculus of the rat. *Brain Research Bulletin.*, *23*, 25-35.

[53] Beart, P. M., Nicolopoulos, L. S., West, D. C. & Headley, P. M. (1998). An excitatory amino acid projection from ventromedial hypothalamus to periaqueductal gray in the rat: autoradiographic and electrophysiology evidence. *Neuroscience Letters.*, *85*, 205-11.

[54] Hrabovszky, E., Halasz, J., Meelis, W., Kruk, M. R, Liposits, Z. & Haller, J. (2005). Neurochemical characterization of hypothalamic neurons involved in attack behavior: glutamatergic dominance and co-expression of thyrotropin-releasing hormone in a subset of glutamatergic neurons. *Neuroscience.*, *133*, 657-66.

[55] Shaikh, M. B., Schubert, K. & Siegel, A. (1994). Basal amygdaloid facilitation of midbrain periaqueductal gray elicited defensive rage behavior in the cat is mediated through NMDA receptors. *Brain Research.*, *635*, 187-95.

[56] Duncan, G. E., Moy, S. S. & Perez, A. (2004). Deficits in sensorimotor gating and tests of social behavior in a genetic model of reduced NMDA receptor function. *Behaviorial Brain Research.*, *153*, 507-19.

[57] Davidson, R. J., Putnam, K. M. & Larson, C. L. (2000). Dysfunction in the Neural Circuitry of Emotion Regulation - Possible Prelude to Violence. *Science.*, 591-594.

[58] Lee, J. J. & Croucher, M. J. (2003). Actions of group I and group II metabotropic glutamate receptor ligands on 5-hydroxytryptamine release in the rat cerebral cortex in vivo: differential roles in the regulation of central serotonergic neurotransmission. *Neuroscience.*, *117*(3), 671-679.

[59] Maione, S., Palazzo, E., de Novellis, V., Stella, L., Leyva, J. & Rossi, F. (1998). Metabotropic glutamate receptors modulate serotonin release in the rat periaqueductal gray matter. *Naunyn Schmiedeberg Arch Pharmacol.*, *358*(4), 411-417.

[60] Lee, H. S., Kim, M. A., Valentino, R. J. & Waterhouse, B. D. (2003). Glutamatergic afferent projections to the dorsal raphe nucleus of the rat. *Brain Res.*, *963*(1-2), 57-71.

[61] Tao, R. & Auerbach, S. B. (2003). Influence of inhibitory and excitatory inputs on serotonin efflux differs in the dorsal and median raphe nuclei. *Brain Res.*, *961*(1), 109-120.

[62] Amat, J., Baratta, M. V., Paul, E., Bland, S. T., Watkins, L. R. & Maier, S. F. (2005). Medial prefrontal cortex determines how stressor controllability affects behavior and dorsal raphe nucleus. *Nat Neurosci.*, *8*, 365- 371.

[63] Takahashi, A., Quadros, I. M., de Almeida, R. M. & Miczek, K. A. (2012). Behavioral and pharmacogenetics of aggressive behavior. *Curr Top Behav Neurosci.*, *12*, 73-138.

[64] Aghajanian, G. K. & Wang, R. Y. (1977). Habenular and other midbrain raphe afferents demonstrated by a modified retrograde tracing technique. *Brain Res.*, *122*, 229-242.

[65] Behzadi, G., Kalen, P., Parvopassu, F. & Wiklund, L. (1990). Afferents to the median raphe nucleus of the rat: retrograde cholera toxin and wheat germ conjugated horseradish peroxidase tracing, and selective d-[3H] aspartate labelling of possible excitatory amino acid inputs. *Neuroscience.*, *37*, 77-100.

[66] Celada, P., Puig, M. V., Casanovas, J. M., Guillazo, G. & Artigas, F. (2001). Control of dorsal raphe serotonergic neurons by the medial prefrontal cortex: Involvement of serotonin-1A, GABAA, and glutamate receptors. *J Neurosci.*, *21* 9917-9929.

[67] Tao, R. & Auerbach, S. B. (2000). Regulation of serotonin release by GABA and excitatory amino acids. *J Psychopharmacol.*, *14*,100-113.

[68] Miczek, K. A. & Fish, E. W. (2006). Monoamines, GABA, Glutamate, and Aggression. In: Nelson, RJ., editor. Biology of aggression. New York: Oxford UP,114-149.

[69] Rewerski, W., Kostowski, W., Piechocki, T. & Rylski, M. (1971). The effects of some hallucinogens on aggressiveness of mice and rats. I. *Pharmacology.*, 1971. 5, 314-320.

[70] Burkhalter, J. E. & Balster, R. L. (1979). The effects of phencyclidine on isolation-induced aggression in mice. *Psychol Rep.*, *45*, 571-576.

[71] Wilmot, C. A., Vander-Wende, C. & Spoerlein, M. T. (1987). The effects of phencyclidine on fighting in differentially housed mice. *Pharmacol Biochem Behav.*, *28*, 341-346.

[72] Krsiak, M. (1974). Behavioral changes andn aggressivity evoked by drugs in mice. *Res Commun Chem Pathol Pharmacol.*, *7*, 237-257.

[73] McAllister, K. H. (1990). Ethological analysis of the effects of MK-801 upon aggressive male mice: similarity to chlordiazepoxide. *Pharmacol Biochem Behav.*, *37*(1), 101-106.

[74] Musty, R. E. & Consroe, P. F. (1982). Phencyclidine produces aggressive behavior in rapid eye movement sleep deprived rats. *Life Sci.*, *30*, 1733-1738.

[75] Tyler, C. B. & Miczek, K. A. (1982). Effects of phencyclidine on aggressive behavior in mice. *Pharmacol Biochem Behav.*, *17*, 503-510.

[76] Lang, A., Harro, J., Soosaar, A. et al. (1995). Role of *N*-methyl-D-aspartic acid and cholecystokinin receptors in apomorphine-induced aggressive behaviour in rats. *Naunyn Schmiedeberg Arch Pharmacol.*, *351*(4), 363-370.

[77] Belozertseva, I. V. & Bespalov, A. Y. (1999). Effects of NMDA receptor channel blockade on aggression in isolated male mice. *Aggress Behav.*, *25*, 381-396.

[78] Miczek, K. A. & O'Donnell, J. M. (1978). Intruder-evoked aggression in isolated and nonisolated mice: Effects of psychomotor stimulants and L-dopa. *Psychopharmacology.*, *57*, 47-55.

[79] Schmidt, W. J. & Reith, M. E. A. (2005). Dopamin and glutamate in psychiatric disorders. Chapter 11, Miczek KA, Fish EW (Eds): Dopamine, Glutamate, and Aggression. Humana Press Inc., Totowa, NJ.

[80] Newman, E. L., Chu, A., Bahamón, B., Takahashi, A., Debold, J. F. & Miczek, K. A. (2012). NMDA receptor antagonism: escalation of aggressive behavior in alcohol-drinking mice. *Psychopharmacology (Berl).*, *224*(1), 167-77.

[81] Cheng, J., Xiong, Z., Duffney, L. J., Wei, J., Liu, A., Liu, S., Chen, G. J. & Yan, Z. (2014). Methylphenidate exerts dose-dependent effects on glutamate receptors and behaviors. *Biol Psychiatry.*, *76*(12), 953-62.

[82] Mathe, J. M., Nomikos, G. G., Schilstrom, B. & Svensson, T. H. (1998). Non-NMDA excitatory amino acid receptors in the ventral tegmental area mediate systemic dizocilpine (MK-801) induced hyperlocomotion and dopamine release in the nucleus accumbens. *J Neurosci Res.*, *51*(5), 583-592.

[83] Lorrain, D. S., Baccei, C. S., Bristow, L. J., Anderson, J. J. & Varney, M. A. (2003). Effects of ketamine and Nmethyl- D-aspartate on glutamate and dopamine release in the rat prefrontal cortex: modulation by a group II selective metabotropic glutamate receptor agonist LY379268. *Neuroscience.*, *117*(3), 697-706.

[84] Takahata, R. & Moghaddam, B. (2003). Activation of glutamate neurotransmission in the prefrontal cortex sustains the motoric and dopaminergic effects of phencyclidine. *Neuropsychopharmacology.*, *28*(6), 1117-1124.

[85] Spanagel, R., Eilbacher, B. & Wilke, R. (1994). Memantine-induced dopamine release in the prefrontal cortex and striatum of the rat—a pharmacokinetic microdialysis study. *Eur J Pharmacol.*, *262*(1-2), 21-26.

[86] Hertel, P., Mathe, J. M., Nomikos, G. G., Iurlo, M., Mathe, A. A. & Svensson, T. H. (1995). Effects of D-amphetamine and phencyclidine on behavior and extracellular

concentrations of neurotensin and dopamine in the ventral striatum and the medial prefrontal cortex of the rat. *Behav Brain Res.*, *72*(1-2), 103-114.

[87] Kuroki, T., Kawahara, T., Yonezawa, Y. & Tashiro, N. (1999). Effects of the serotonin2A/2C receptor agonist and antagonist on phencyclidine-induced dopamine release in rat medial prefrontal cortex. *Prog Neuropsychopharmacol Biol Psychiatry.*, *23*(7), 1259-1275.

[88] Siegel, A., Roeling, T. A. P., Gregg, T. R. & Kruk, M. R. (1999). Neuropharmacology of brain-stimulationevoked aggression. *Neurosci Biobehav Rev.*, *23*, 359-389.

[89] Bandler, R. (1984). Identification and midbrain periaqueductal grey neurones mediating aggressive andvdefensive behavior by intracerebral microinjections of excitatory amino acids. In: Bandler, R., (ed). Modulation of sensorimotor activity during alterations in behavior states. *New York: Alan R. Liss, Inc*, 369-91.

[90] Lu, C. L., Shaikh, M. B. & Siegel, A. (1992). Role of NMDA receptors in hypothalamic facilitation of feline defensive rage elicited from the midbrain periaqueductal gray. *Brain Research.*, *581*, 123-32.

[91] Schubert, K., Shaikh, M. B. & Siegel, A. (1996). NMDA receptors in the midbrain periaqueductal gray mediate hypothalamically evoked hissing behavior in the cat. *Brain Research.*, *726*, 80-90.

[92] Vekovischeva, O. Y., Aitta-aho, T., Echenko, O., Kankaanpaa, A., Seppala, T., Honkanen, A. et al. (2004). Reduced aggression in AMPA-type glutamate receptor GluR-A subunit-deficient mice. *Genes, Brain, and Behavior.*, *3*, 253-65.

[93] Vekovischeva, O. Y., Aitta-aho, T., Verbitskaya, E., Sandnabba, K. & Korpi, E. R. (2007). Acute effects of AMPA-type glutamate receptor antagonists on intermale social behavior in two mouse lines bidirectionally selected for offensive aggression. *Pharmacology, Biochemistry, and Behavior.*, *87*, 241-9.

[94] Adamczyk, A., Mejias, R., Takamiya, K., Yocum, J., Krasnova, I. N., Calderon, J. et al. (2012). GluA3-deficiency in mice is associated with increased social and aggressive behavior and elevated dopamine in striatum. *Behavioral Brain Research.*, *229*, 265-72.

[95] Brodkin, E. S., Goforth, S. A., Keene, A. H., Fossella, J. A. & Silver, L. M. (2002). Identification of quantitative trait loci that affect aggressive behavior in mice. *Journal of Neuroscience.*, *22*, 1165-70.

[96] Navarro, J. F., de Castro, V. & Martin-Lopez, M. (2008). JNJ16259685, a selective mGlu(1) antagonist, suppresses isolation-induced aggression in male mice. *European Journal of Pharmacology.*, *586*, 217-20.

[97] Navarro, J. F., Postigo, D., Martin, M. & Buron, E. (2006). Antiaggressive effects of MPEP, a selective antagonist of mGlu5 receptors, in agonistic interactions between male mice. *European Journal of Pharmacology.*, *551*, 67-70.

[98] Newman, E. L., Chu, A., Bahamón, B., Takahashi, A., DeBold, J. F. & Miczek, K. A. (2012). NMDA receptor antagonism: escalation of aggressive behavior in alcohol-drinking mice. *Psychopharmacology.*, *224*, 167-177.

[99] Ago, Y., Araki, R., Yano, K., Kawasaki, T., Chaki, S., Nakazato, A. et al. (2012). The selective metabotropic glutamate 2/3 receptor agonist MGS0028 reverses isolation rearing-induced abnormal behaviors in mice. *Journal of Pharmacological Science.*, *118*, 295-8.

[100] Navarro, J. F., de Castro, V. & Martin-Lopez, M. (2009). Behavioural profile of selective ligands for mGlu7 and mGlu8 glutamate receptors in agonistic encounters between mice. *Psicothema.*, *21*, 475-9.

[101] Shigemoto, R., Kinoshita, A., Wada, E., Nomura, S., Ohishi, H., Takada, M., Flor, P. J., Neki, A., Abe, T., Nakanishi, S. & Mizuno, N. (1997). Differential presynaptic localization of metabotropic glutamate receptor subtypes in the rat hippocampus. *J Neurosci.*, *17*, 7503-7522.

[102] Shaltiel, G., Maeng, S., Malkesman, O., Pearson, B., Schloesser, R. J., Tragon, T., Rogawski, M., Gasior, M., Luckenbaugh, D., Chen, G. & Manji, H. K. (2008). Evidence for the involvement of the kainate receptor subunit GluR6 (GRIK2) in mediating behavioral displays related to behavioral symptoms of mania. *Mol Psychiatry.*, *13*, 858-872.

[103] Yadav, R., Gupta, S. C., Hillman, B. G., Bhatt, J. M., Stairs, D. J. & Dravid, S. M. (2012)- Deletion of glutamate delta-1 receptor in mouse leads to aberrant emotional and social behaviors. *PloS One.*, *7*. e32969.

[104] Monroe, R. R. (1978). Brain Dysfunction in Aggressive Criminals. Lexington: DC Heath and Company,

[105] Siegel, A. & Mirsky, A. F. (1994). The neurobiology of violence and aggression. In: Reiss AJ, Jr., ed. Understanding and Preventing Violence. Biobehavioral Influences. Washington, DC: National Academy Press, 59-172.

[106] Raine, A., Buchsbaum, M. S., Stanley, J., Lottenberg, S., Abel, L. & Stoddard, J. (1994). Selective reductions in prefrontal glucose metabolism in murderers. *Biol Psychiatry.*, *36*(6), 365-373.

[107] Raine, A., Lencz, T., Bihrle, S., LaCasse, L. & Colletti, P. (2000). Reduced prefrontal gray matter volume and reduced autonomic activity in antisocial personality disorder. *Arch Gen Psychiatry.*, *57*(2), 119-127.

[108] Best, M., Williams, J. M. & Coccaro, E. F. (2002). Evidence for a dysfunctional prefrontal circuit in patients with an impulsive aggressive disorder. *Proc Natl Acad Sci USA.*, *99*(12), 8448-8453.

[109] Juhasz, C., Behen, M. E., Muzik, O., Chugani, D. C. & Chugani, H. T. (2001). Bilateral medial prefrontal and temporal neocortical hypometabolism in children with epilepsy and aggression. *Epilepsia.*, *42*(8), 991-1001.

[110] King, B. H., Wright, D. M., Handen, B. L. et al. (2001). Double-blind, placebo-controlled study of amantadine hydrochloride in the treatment of children with autistic disorder. *J Am Acad Child Adolesc Psychiatry.*, *40*(6), 658-665.

[111] Coccaro, E. F., Lee, R. & Vezina, P. (2013). Cerebrospinal fluid glutamate concentration correlates with impulsive aggression in human subjects. *J Psychiatr Res.*, *47*(9), 1247-53.

[112] Faustman, W. O., Bardgett, M., Faull, K. F., Pfefferbaum, A. & Csernansky, J. G. (1999). Cerebrospinal fluid glutamate inversely correlates with positive symptom severity in unmedicated male schizophrenic/ schizoaffective patients. *Biological Psychiatry.*, *45*, 68-75.

[113] Gattaz, W. F., Gasser, T. & Beckmann, H. (1985). Multidimensional analysis of the concentrations of 17 substances in the CSF of schizophrenics and controls. *Biological Psychiatry.*, *20*, 360-6.

[114] Do, K. Q., Lauer, C. J., Schreiber, W., Zollinger, M., Gutteck-Amsler, U., Cuenod, M. et al. (1995) γ- Glutamylglutamine and taurine concentrations are decreased in the cerebro-spinal fluid of drugnaive patients with schizophrenic disorders. *Journal of Neurochemistry.*, *65*, 2652-62.

[115] de la Fuente-Sandoval, C., Leon-Ortiz, P., Favila, R., Stephano, S., Mamo, D., Ramirez-Bermudez, J. et al. (2011). Higher levels of glutamate in the associative-striatum of subjects with prodromal symptoms of schizophrenia and patients with first-episode psychosis. *Neuropsychopharmacology.*, *36*(9), 1781-91.

[116] Frye, M. A., Tsai, G. E., Huggins, T., Coyle, J. T. & Post, R. M. (2007). Low cerebrospinal fluid glutamate and glycine in refractory affective disorder. *Biological Psychiatry.*, *61*, 162-6.

[117] Frye, M. A., Watzl, J., Banakar, S., O'Neill, J., Mintz, J., Davanzo, P. et al. (2007). Increased anterior cingulate/medial prefrontal cortical glutamate and creatine in bipolar depression. *Neuropsychopharmacology.*, *32*, 2490-9.

[118] Chakrabarty, K., Bhattacharyya, S., Christopher, R. & Khanna, S. (2005). Glutamatergic dysfunction in OCD. *Neuropsychopharmacology.*, *30*, 1735-40.

[119] Ballard, C. G., Gauthier, S., Cummings, J. L., Brodaty, H., Grossberg, G. T., Robert, P. & Lyketsos, C. G., (2009). Management of agitation and aggression associated with Alzheimer disease. *Nat Rev Neurol.*, *5*, 245-255.

[120] Wilcock, G. K., Ballard, C. G., Cooper, J. A. & Loft, H. (2008). Memantine for agitation/aggression and psychosis in moderately severe to severe Alzheimer's disease: a pooled analysis of 3 studies. *J Clin Psychiatry.*, *69*, 341-348.

[121] Hamshere, M. L., Green, E. K., Jones, I. R., Jones, L., Moskvina, V., Kirov, G., Grozeva, D., Nikolov, I., Vukcevic, D., Caesar, S., Gordon-Smith, K., Fraser, C., Russell, E., Breen, G., St Clair, D, Collier, D. A., Young, A. H., Ferrier, I. N., Farmer, A., McGuffin, P., Holmans, P. A., Owen, M. J, O'Donovan, M. C. & Craddock, N. Wellcome Trust Case Control Consortium. (2009). Genetic utility of broadly defined bipolar schizoaffective disorder as a diagnostic concept. *Brit J Psychiat.*, *195*, 23-29.

[122] Carrillo, M., Ricci L. A. & Melloni R. H. (2011). Glutamate vasopressin interactions and the neurobiology of anabolic steroid induced offensive aggression. *Neuroscience.*, *185*, 85-96.

[123] Zalcman S. S. & Siegel A. (2006.). The neurobiology of aggression and rage: role of cytokines. *Brain Behav Immun.*, *20*(6), 507-14.

[124] Baranzini S. E., Srinivasan R., Khankhanian P., Okuda D. T., Nelson S. J., Matthews P. M., Hauser S. L., Oksenberg J. R. & Pelletier D. (2010). Genetic variation influences glutamate concentrations in brains of patients with multiple sclerosis. *Brain.*, *133*(9), 2603-11.

[125] Hamshere M. L, Green E. K., Jones I. R., Jones L., Moskvina V., Kirov G., Grozeva D., Nikolov I., Vukcevic D., Caesar S., Gordon-Smith K., Fraser C., Russell E., Breen G., St Clair D, Collier D. A., Young A. H., Ferrier I. N., Farmer A., McGuffin P., Wellcome Trust Case Control Consortium, Holmans P. A., Owen M. J., O'Donovan M. C. & Craddock N. (2009). Genetic utility of broadly defined bipolar schizoaffective disorder as a diagnostic concept. *Br J Psychiatry.*, *195*(1), 23-9.

In: Modulators of Glutamatergic Signaling
Editor: Zoran M. Pavlovic

ISBN: 978-1-63483-493-3

Chapter 2

SHOULD WE BE EXCITED ABOUT GLUTAMATE DYSREGULATION IN THE ETIOLOGY OF ADHD? A REVIEW OF THE DATA

S. R. Batten[1], G. A. Gerhardt[1] and P. E. A. Glaser[2,*]
[1]Department of Anatomy and Neurobiology, University of Kentucky, KY, US
[2]Department of Psychiatry, Washington University in St. Louis, MO, US

ABSTRACT

In this chapter we review current data assessing the role of glutamate in the etiology of ADHD. A general introduction of ADHD as well as common comorbidities are briefly discussed. The glutamate system in general as well as its potential role in ADHD is thoroughly reviewed evaluating both preclinical and clinical data. The current ADHD treatments that act on the glutamate system, memantine and atomoxetine, are discussed. The chapter concludes with a discussion on the future of glutamatergic drugs in the treatment of ADHD.

Keywords: ADHD, brain energetics, glutamate, glutamatergic pharmacotherapies

INTRODUCTION

Attention-deficit hyperactivity disorder (ADHD) is a disorder commonly diagnosed in children and adolescents that begins in childhood and often continue into adulthood [1, 2]. ADHD is considered a neurodevelopmental disorder that is characterized by deficits in attention and impulse control, often accompanied by hyperactivity [1, 3]. Currently, the etiological processes of ADHD are not well understood despite the fact that it affects 5% of children and 2.5% of adults worldwide [4].

* Corresponding author: Department of Psychiatry, Washington Univ. in St. Louis 660 South Euclid Campus Box #8134, St. Louis, Mo 63110, Email: glaserp@psychiatry.wustl.edu.

Historically speaking, the earliest reference of ADHD or a 'disease of attention' is rumored to be by Shakespeare in his play *King Henry VII* [1]. However, a better-documented report was in 1902 by the English physician, George Still, who focused on clinical aspects of those with ADHD-like symptoms and argued that there were certain traits similar in all those who seemed to have this 'disorder' [5]. Through the years ADHD has had several different nomenclatures and diagnostic criteria [1]. Currently, the DSM-V denotes three presentations of ADHD including inattentive (ADHD-PI), hyper-active/impulsive (ADHD-HI), and combined (ADHD-C) [4]. The DSM-V also established more 'adult relevant' symptoms as well as the classification system for severity as mild, moderate, and severe [4].

Regardless of the nomenclature used throughout the years, it has been hypothesized for decades that ADHD is a heterogeneous disorder involving interactions between genetics and the environment [6, 7]. Environmentally speaking, positive correlations are observed between ADHD and lower socioeconomic status, parent criminality, parental mental health, fetal exposure to alcohol, and maternal smoking as well as many other factors [1, 8, 9]. From a genetic standpoint most genes associated with ADHD seem to be related to malfunctions in several neurotransmitter systems with the most studied being those associated with the dopaminergic, noradrenergic, and serotonergic systems [9, 10, 11, 12].

More recently, evidence suggests that genetic and neurobiological issues related to the glutamate system may be a contributing factor in the etiology of ADHD [13, 14, 15, 16]. Here we will outline comorbidities associated with ADHD as well as discuss the glutamate system in general, review data supporting the role of glutamate in ADHD, discuss current glutamatergic pharmacotherapies, and discuss future perspectives on glutamatergic drugs in meliorating the symptoms of ADHD.

ADHD Comorbidities

In a study assessing children and adolescents referred for ADHD, comorbidity ranged from ten to fifty percent for other common psychiatric disorders of youth [17]. Another study found roughly 62% of ADHD patients are likely to have one or more comorbidities and 35% of patients are likely to have two or more psychiatric comorbidities [18]. However, no differences in comorbidities seem to exist between males and females diagnosed with ADHD [18].

The main psychiatric conditions that accompany ADHD include disruptive behavior disorders (30-50%), depression (15-20%), anxiety disorders (25%), learning disabilities (10-15%), and substance abuse disorders (9-40%) [19]. Of the disruptive disorders, a strong association is seen with ODD/CD [18]. Also, those with comorbid ADHD and major depression are more likely to develop bipolar disorder later in life [20].

There is no difference in comorbidity rate between males and females with ADHD however there are differences in the type of diseases comorbid with ADHD between the sexes [18]. Specifically, males are more likely to have externalizing-only disorders (i.e., ODD/CD) and females internalizing-only disorders (i.e., anxiety) [18]. Thus, ADHD is a disorder that is likely to be accompanied by a myriad of other disorders. These comorbid data should be considered when thinking of the glutamatergic system's role in ADHD and how the glutamate system may promote other psychiatric disorders.

ADHD and the Glutatmate System

Several studies have shown issues with the prefrontal cortex (PFC) as well as other 'higher level' circuits in the brain in those with ADHD [21-23]. Considering the many glutamatergic projections to the PFC, as well as to other cortical regions, it is no surprise that researchers have begun exploring the glutamate system in those with ADHD [23]. Glutamatergic changes in human ADHD have also stimulated preclinical work on glutamate and ADHD [14, 24]. Thus, it is becoming apparent that dysregulations in the glutamate system may be one key etiologic factor in ADHD. In this section we will examine general aspects of the glutamate system as well as potential glutamatergic dysfunctions seen in ADHD.

Glutamate System Basics

Glutamate is the major excitatory neurotransmitter in the CNS [25]. Glutamate is synthesized in the nerve terminals of glutamatergic neurons from two major sources: α-ketoglutarate produced from the TCA cycle or from glutamine shuttled to neurons from glial cells [26-28]. In neuron terminals α-ketoglutarate and glutamine are enzymatically synthesized into glutamate and packaged into synaptic vesicles in an energy dependent fashion via vesicular glutamate transporters (VGLUT) [29].

Once packaged, glutamate can then be released into the synaptic cleft, in a Ca^{2+} dependent manner, upon the firing of an action potential [30, 31]. Once glutamate is released it is free to: (1) bind to pre and post-synaptic receptors, (2) be taken up by glial cells in a Na^+ dependent fashion, (3) be actively transported into presynaptic neurons and repackaged, (4) diffuse away from the synapse [27, 28, 32].

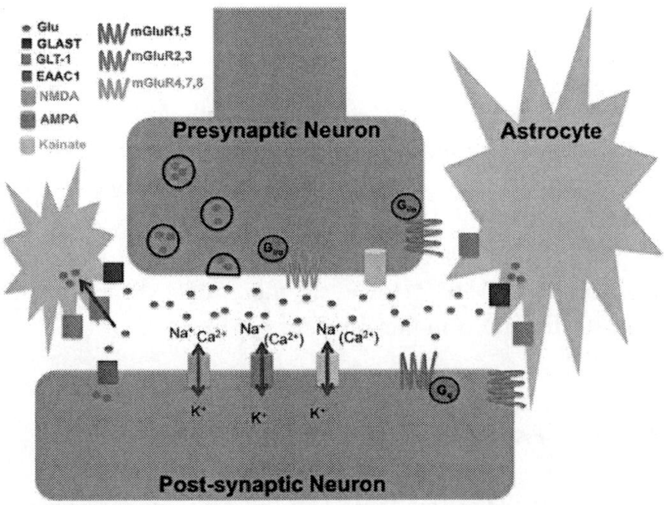

Figure 1. Glutamatergic synapse simplified. Once glutamate is released from the presynaptic neuron it is free to bind pre and post ionotropic and metabotropic receptors or be taken up by high affinity EAATs on glia cells in a Na^+ dependent fashion. Binding to AMPA, NMDA, Kainate, or mGluR1, 5 receptors will cause excitation. Binding to mGluR2, 3, 4, 7, 8 receptors causes cellular inhibition.

In the synaptic cleft glutamate may bind to ionotropic receptors (NMDA, AMPA, Kainate), presynaptic inhibitory metabotropic receptors (mGluR2, 3, 4, 7, 8), and/or post-synaptic excitatory metabotropic receptors (mGluR1, 5) [30, 33, 34]. Approximately 90% of glutamate is taken up by astrocytes either by excitatory amino acid transporter 1 or 2 (EAAT 1 and EAAT2 respectively) [33-35]. Glutamate can also be taken up by neurons via EAAT3-5 although this seems to be a secondary process for this neurotransmitter system [33-35]. See Figure 1 for a simplified diagram of the glutamate synapse.

Bridging Old and New Thoughts about Neural Signaling in ADHD

Researchers became further interested in glutamate dynamics in the PFC in those with ADHD after realizing the reciprocal modulatory nature between these glutamatergic circuits and the dopamine system (the system historically most widely studied in ADHD) [14]. Specifically, glutamate projections from the PFC project to dopamine rich areas such as the striatum, nucleus accumbens, ventral tegmental area (VTA), and substantia nigra [33, 37]. Dopamine projections from the VTA and nucleus accumbens also project to the PFC [33, 37]. Thus, malfunctions in either of these systems could cause dysregulations in in the other [14].

Studies have shown that the NMDA receptor is crucial in stimulating dopamine neurons in the VTA and substantia nigra [38, 39]. Also, there is evidence that D2 receptor activation may inhibit the excitatory effects of NMDA receptors [40]. Similarly, D4 receptor activation may decrease the excitatory response of AMPA receptors in PFC pyramidal neurons [41]. Thus, glutamatergic dysregulations in ADHD may be as detrimental as those seen in the dopamine system and therefore normalization of this system may help alleviate symptoms seen in ADHD.

ADHD and Glutamate: Evidence from Human Studies

Decreased functioning of the PFC and subsequent deficits in cognitive functioning including working memory has been observed in those with ADHD [42, 43]. An MRI study has shown that there is an increase marker for glutamate in the anterior cingulate cortex of those with ADHD compared to controls [23]. There is also evidence that treating children with medications know to meliorate the symptoms of ADHD decreases glutamate levels in the striatum and PFC [44]. While this research is promising, there is not robust clinical data on the role of glutamate in ADHD. However, more data from preclinical models are suggestive of glutamate dysfunctions in this disorder.

ADHD and Glutamate: Evidence from Animal Research

The spontaneously hypertensive rat (SHR) is the most widely used model of ADHD specifically for studying ADHD combined type [45, 46]. The SHR shows attention deficits, hyperactivity, and impulsivity in motor movements [47-49]. While this model has been widely used in studying ADHD pathology there are recent arguments about how well this animal model translates to the human disease, what animal models may represent other subtypes, and what the appropriate control should be in experiments [50]. That being said, most agree that the SHR from Charles River (SHR/NCrl) is the most appropriate animals model for ADHD-C, the Winstar-Kyoto from Charles River (WKY/NCrl) is best suited as a model for ADHD-PI (predominately inattentive type), and that the WKY from Harlan (WKY/NHsd) is the most appropriate control with the outbred Sprague-Dawley (SD) being another potential control [46, 51-53]. Several different studies have shown glutamate dysregulation in these animals.

Studying glutamate dynamics, specifically in the PFC, seems to be particularly important in understanding the etiology of ADHD [21-23]. Evidence from our lab has shown a significant increase in KCl-evoked glutamate release in regions of the PFC, the cingulate and infralimbic cortices, in the SHR compared to the WKY [24]. It is worth noting that evidence suggests that the cingulate cortex may regulate emotions in humans as well as be a primary center for motivation [54, 55]. On the other hand the infralimbic region seems to be related to attentional focus as well as attentional set-shifting [56]. Note that the defecits with tasks related to the aforementioned brain areas are often seen in ADHD [57-59].

Data collected also suggest that glutamate release may be increased in the striatum of the SHR compared to the WKY [24]. This finding may relate to ADHD in that the striatum is related to movement and reward circuitry both of which seem to be disrupted in ADHD [60, 61]. It also takes a significantly higher volume of ejected glutamate to achieve similar peak amplitudes in SHR rats compared to WKY control suggesting uptake may be faster in the PFC of SHRs; this may be a potential compensatory response to increased vesicular glutamate release in this area [24]. Considering the differences seen above in anesthetized rats our lab has also attempted to assess glutamate dynamics in freely-moving animals.

Evidence suggests that tonic glutamate concentrations are higher in SHR animals compared to WKY rats regardless of treatment with methylphenidate [24]. Specifically, higher tonic levels were seen in the cingulate, prelimbic, and infralimbic cortices compared to WKY [24]. Similarly, tonic glutamate levels are increased in the PFC of both intermediate and chronically treated methylphenidate animals compared to SHR saline controls [24]. Further, phasic glutamate was decreased in SHR rats treated chronically with methylphenidate compared to SHR rats treated with saline [24].

It is worth noting that the type of phasic signal seen in the animals in this experiment differed between strain and treatment [24]. Specifically, SHR's treated intermediately with methylphenidate had more rapid, multi-peaks compared with WKY's treated with methylphenidate [24]. However, when treated chronically with methylphenidate, the SHR rats had more slow phasic events compared to SHR saline controls [24]. Overall, these freely-moving data suggest that there are dynamic changes in PFC tonic and phasic glutamate levels in the SHR treated with a well know dopamine acting drug often prescribed to those with ADHD.

More 'indirect measures' have also shown differences in glutamate signaling in the SHR. For example, there is evidence that NMDA receptor activation resulted in less calcium influx in SHR PFC slices compared to those obtained from WKY [62]. Glutamate applied to SHR PFC slices also showed an increase in norepinephrine release by activation of NMDA receptors compared to WKY controls [63].

Increased glutamate levels have also been shown to stimulate release of dopamine in the substantia nigra of SHR rats compared to WKY controls suggesting that there may be altered regulation of dopamine by glutamate in the SHR [39]. Further, mice with inactivation of the dopamine transporter showed increased hyperactivity when administered NMDA antagonists and the hyperactivity decreased when given drugs that increased glutamate signaling [64, 65]. Neonatal 6-hydroxydopamine lesions in rats caused dose dependent decreases in D4 receptors and increases in glutamate transporters in the striatum [66]. All of this data further suggests tightly coupled interactions between the glutamate and dopamine systems in ADHD [66]

Inactivation of mGluR5 increases hyperactivity in mice [67]. Further, impairment of presynaptic mGLuR7 receptors using MMPIP in the prelimbic cortex decreased visuospatial attention [68]. However, in this same study no other NMDA antagonists used or mGluR2/3 inhibitors caused any changes in impulse control when infused into the prelimbic or infralimbic areas [68]. Nevertheless, considering the above, mGluR drugs may work to normalize glutamatergic signaling. Namely, agonist of mGluR receptors may normalize the glutamatergic system and behavior in these animals [69]. Along with this, downstream effectors of mGluRs may be dysregulated in ADHD as well.

Ca^{2+}/calmodulin-dependent protein kinase II (CaMKII) and protein kinase C (PKC) signaling in the PFC in glutamatergic neurons may contribute to ADHD symptomology. CaMKII and PKC are essential for long-term potentiation (LTP); a process highly associated with glutamatergic signaling. In fact it was found that increase in CaMKII autophosphorylation and GluR1 phosphorylation was found in the mPFC of stroke prone spontaneous hypertensive rats (SHRSP) compared to WKY controls suggesting mPFC pathology. This mPFC pathology was realated to impaired performance on both the Y maze task and a novel discrimination task [70]. Furthermore, methylphenidate treatment not only increased SHRSP performance on behavioral tasks but also decreased CaMKII autophosphorylation and GluR1 phosphorylation in the mPFC compared to WKY controls [70].

Using the 5-choice serial reaction time task (5CSRTT) it has been reported that NMDA antagonism by MK-801 in the infralimbic cortex increased impulsive responding [68]. In another study assessing the role of the glutamate system in impulsive choice (delay and probability discounting) it was found that systemic injections of MK-801 decreased discounting rate for the larger reinforcer (less impulsive) [71]. These discrepancies may be due to the different underlying processes thought to contribute to impulsive action compared to impulsive choice [72]. Regardless, these data do suggest that some kind of interaction is occurring with the glutamate system in models of impulsivity although from these studies the directionality is not clear.

Glutamate, ADHD, and Genetic Correlates

Latrophilin-3 (LRHN3), an adhesion G-protein coupled-receptor indicated in synaptogenesis and synaptic plasticity, has been shown to promote ADHD like behaviors in experimental models with loss-of-function mutation in the *LRHN3* gene [73]. Specifically, LRHN3 knockdown causes increased locomotor activity and dopamine signaling and can be rescued by drugs used to treat ADHD [74].

Fibronectin (FLRT3) is a transmembrane protein and a natural ligand of LRHN3; this binding is essential to glutamatergic signaling [75]. When FLRT3 and LRHN3 bind this action seems to regulate excitatory synapses and plasticity both pre and post-synaptically

[76]. Some with ADHD have been shown to contain mutations in FLRT3 further suggesting links between glutamate signaling and ADHD [77].

Mutations resulting in malfunctioned SHANK proteins may also cause ADHD symptoms [78]. SHANK proteins are synaptic multi-domain scaffold proteins of the post-synaptic density that connect receptors, ion channels, and other types of membrane proteins to actin cytoskeleton G protein-coupled pathways involved in dendrite maturation and synapse formation [79]. Gene mutations in nitric oxide synthase-1 (NOS-1), a protein closely associated with the NMDA receptor and responsible for nitric oxide generation, have been linked to ADHD behaviors and impulsivity as well [16, 80].

Mutations in several other genes including those that code for NMDA receptor subunit-2A and 2B as well as genes encoding for glial glutamate transporter EAAT1 and mGluRs have all been shown to be associated with ADHD [79, 81-83].

Brain Energetics and Glutamate: A New Way to View ADHD?

The brain is an interesting organ in the fact that it is about 2% of a human's mass but utilizes approximately 20% of its oxygen and glucose [84]. Further, unlike other organs in the body the brain can only use sugars for energy due to the blood-brain barrier [85]. Neurons are also interesting in the fact that they are mercurial in nature; they can go from quiescence to rapid firing in seconds [86]. Even more interesting is that neurons cannot store their own energy; this is a task left to glia that store energy in the form of glycogen [87]. While these aforementioned facts are generally accepted the molecule used as the primary energy source in the brain is still debated [88].

It is a well-accepted scientific fact that most of the body uses glucose as its primary energy source. While the brain does seem to use glucose to some degree evidence suggest that lactate may be used more frequently as an energy source by the brain [88]. The balance and shuttling of glucose and lactate in the brain is tightly performed and regulated by neuron-glia coupling [89].

Glucose is transported into the brain through blood vessels and into astrocytes via glucose transporters (GLUT1) [89]. Glucose is then transformed into glucose-6-phosphate then either enzymatically transformed into glycogen via glycogen synthase or turned into pyruvate then lactate via pyruvate kinase and lactate dehydrogenase respectively [88, 89]. Lactate can then be shuttled out of astrocytes via monocarboxylate transporter 4 (MCT4) and shuttled into neurons via MCT2 [88, 89]. Once lactate is in neurons it is then transformed back into pyruvate and taken up by mitochondria for use in the TCA cycle (Figure 2) [88, 89].

It has been shown that there is tight coupling between glutamate signaling and lactate shuttling [89]. Evidence suggests that when glutamate is release from neurons and taken up by glia via EAATs that this Na^+ dependent process causes lactate to be released to neurons to further support their firing [89]. Also, memory formation, a process highly associated with glutamate, also seems to be associated with lactate [90, 91]. Research shows that blocking lactate transport in the brain may inhibit memory formation [90, 91]. Also, lactate but not glucose could 'rescue' this loss of memory [91]. Further, evidence suggest that beyond lactate's inherent energy properties is may also serve as a second messenger for potentiating LTP likely by regulating the redox state of neurons [92]. Considering this research, there is ample evidence to suggest that the dysregulations in brain energy can directly affect glutamate signaling. Thus, insufficiencies in how the brain acquires energy could directly affect neurotransmitter systems and cause disease.

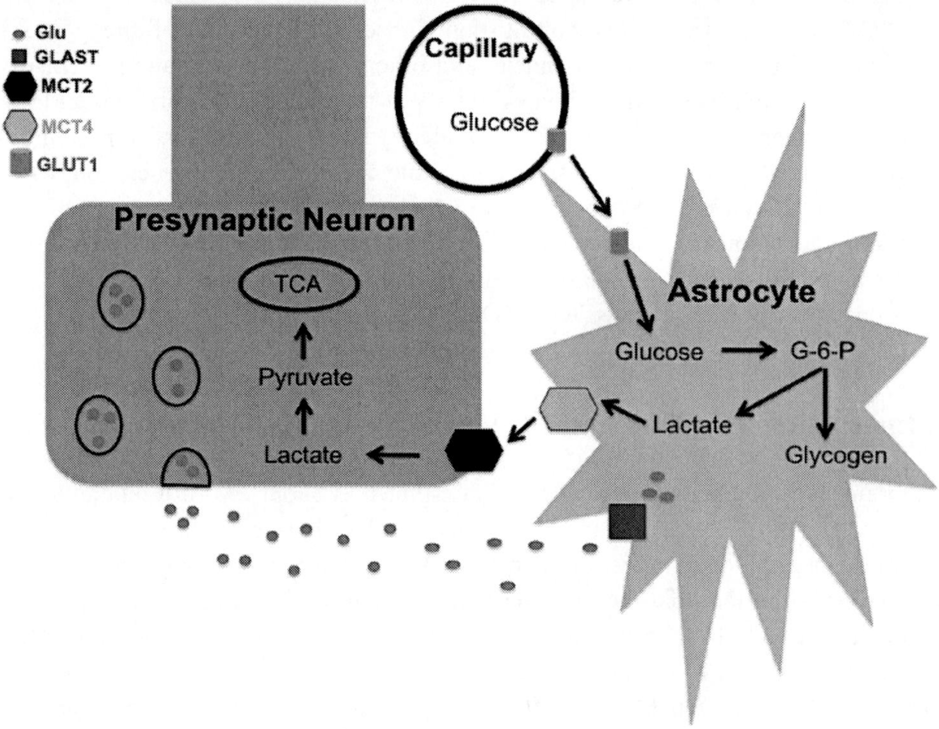

Figure 2. The Astrocyte Neuron Lactate Shuttle. Glucose is transported into the brain through blood vessels and into astrocytes via glucose transporters (GLUT1). Glucose is then transformed into glucose-6-phosphate (G-6-P) then either transformed into glycogen or into lactate. Lactate can then be shuttled out of astrocytes via monocarboxylate transporter 4 (MCT4) and shuttled into neurons via monocarboxylate transporter 2 (MCT2). Once lactate is in neurons it is then transformed into pyruvate and taken up by mitochondria for use in the TCA cycle.

Energetics and astrocytic control of glucose was proposed as a possible etiology for ADHD by a review paper in 2001 [93]. More recently some research has shown that lactate hypo-function in the brain can cause aberrant neuronal firing promoting the symptoms seen in ADHD [94]. In their elegant paper Killeen et al. (2013) proposed the behavior neuroenergetics theory of ADHD. The neuroenergetics theory assumes brain lactate hypo-function in those with ADHD and uses the lactate-neuron shuttle as a 'bottle neck' in neuronal energy acquisition. Using the neuroenergetics theory, with the help of the neuroenergetics mass-action model, these authors derived equations tying neuron physiology to behavior [94].

This model inferred that those with ADHD could only bring approximately 75-80% of their neurocognitive energy to exhibit on tasks [94]. Further, ADHD individuals could only allocate approximately 85% of their cognitive resources to behavioral tasks compared to controls [94]. Even more impressive then the aforementioned was that parameters derived from the model predicted performance on behavioral tasks suggesting that brain lactate levels are predictive of behavioral outcomes [94]. This fact alone suggests hypo-energetics and changes in glutamate signaling may be very important in the etiology of ADHD.

Other researchers have also found a link between brain energetics, neurotransmitter dysregulations, and ADHD. For example, MRI studies suggest that the ability of those with ADHD to 'summon' glucose for oxidative metabolism is impaired in the fronto-parietal system compared to typically developing controls [95]. Further, some data support that those with ADHD may have enzyme deficiencies related to energy supply and that these deficiencies may cause developmental delays in the fronto-striatal circuitry [96]. Evidence also suggests that those with ADHD may require more brain energy than controls to function due to inefficient neural networks lacking full myelination [97].

Dysfunctional GABAergic fibers are also implicated in promoting impulsivity in those with ADHD due to lack these interneurons to inhibit upper motor neurons [98]. This GABAergic dysfunction may be caused by lack of energy supply; this is especially relevant when considering that fast-spiking GABAergic neurons may fail to completely inactivate their sodium channels upon action potential firing thus allowing two times the amount of sodium to enter the cell [99]. This larger intake of sodium then increases the ATP needed for GABA cells to adequately restore membrane potential thus this may making these cells especially vulnerable to energy insufficiencies [94].

N-Acetylaspartate (NAA) levels, a neuro-specific energy storage molecule important for energy metabolism and myelin sheath formation, is altered in ADHD [100-102]. Evidence suggests that treatment with methylphenidate increased NAA levels in the anterior cingulate cortex in those with ADHD [103]. Considering that the anterior cingulate cortex is essential in error processing and learning from mistakes (a deficit seen in those with ADHD) this increase in the energy metabolite NAA may be crucial to ADHD symptoms [104]. The implication of NAA levels in ADHD etiology is even more pronounced when we again consider that neurons in the ADHD brain may be slower processing, more energy insufficient, and variable due to lack of myelination [97, 105, 106].

It is worth noting that there is evidence against lactate as an energy source for neurons, glutamate-lactate coupling, and energy deficits in those with ADHD [107, 108]. However, considering the overwhelming evidence for a relationship between energy dynamics, glutamate, and ADHD it is likely that issues with brain energy may be a primary cause of the disease and perhaps even a potential target of future medications.

TARGETING THE GLUTAMATE SYSTEM TO TREAT ADHD

Recently studies have suggested that not only may the glutamate system promote ADHD pathology but also that targeting the glutamate system with pharmacological agents may normalize the system and alleviate symptoms associated with ADHD [109-111]. It is worth noting that research has currently suggested that staple treatments in ADHD such as methylphenidate and amphetamines may also affect the glutamate system indirectly [14]. However, here we will only discuss current pharmaceuticals approved/suggested to treat ADHD that act directly on the glutamate system namely atomoxetine and memantine. For each drug a brief pharmacological profile will be discussed followed by research from animal models and human studies.

Atomoxetine: Basic Pharmacology

Atomoxetine is a non-stimulant medication associated with less abuse liability than traditional ADHD treatments [112]. Further, data suggest that it is well tolerated with minimal adverse advents in children with ADHD [113]. Initially the primary mechanism of action of atomoxetine was thought to be as a blocker of the norepinephrine transporter (NET) with minimal activity at other neurotransmitter systems [114]. However, more recent evidence suggests that atomoxetine also works as a non-competitive antagonist at NMDA receptors at relevant physiological levels [115, 116]. In children and adolescents atomoxetine was absorbed with peak plasma concentrations occurring in 1-2 hours with a half-life of approximately 3 hours [117]. Generally speaking, the recommended dose of atomoxetine for children is 80 mg/day with a maximum recommended dose of 100 mg/day [118]. Atomoxetine is primarily metabolized by in the liver by CYP2D6; those who are poor metabolizers use the CYP2C19 pathway [118]. Note that it is important to assess for poor metabolism because the drug dose will need to be adjusted accordingly [118].

Atomoxetine and ADHD: Evidence from Animals Models

Several studies have been conducted using atomoxetine in animal models of ADHD and impulsivity. A recent study showed that rats with a lesion in the dorsal noradrenergic ascending bundle given atomoxetine had increased performance on the 5-CSRTT task suggesting that not only can atomoxetine decrease impulsivity but that atomoxetine's effects on the NET are not the sole reason for the drug's pharmacological efficacy [119]. A separate study also found atomoxetine to decrease impulsivity as measured by the 5-CSRTT task [120].

Contrary to the previous studies, Dommett (2014) found that atomoxetine had no effect on SHR performance in the 5-CSRTT. The author of this study suggested that lack of statistical significance could be due to poor sensitivity of the test to adequately tap into the hypothetical construct of impulsivity or that the SHR is not a good model of ADHD [121]. While these explanations cannot be completely ruled out it is worth noting that there is ample evidence to suggest that this may not be the case; however, these topics are discussed elsewhere [47, 122, 123]. It is also worth noting that Dommett (2014) had a difficult time training the SHR on the 5-CSRTT with only ten out of eighteen completing the task. Thus, there may have been an issue in the way the task was set up in this experiment that may have confounded the results.

Another study showed that atomoxetine reversed locomotor activity, impaired novel object recognition, and prepulse inhibition in impulsive mice [124]. A study assessing probability discounting in rats showed that a low dose of atomoxetine increased choice for the larger reinforcer suggesting a decrease in impulsive choice [125]. However, rats in a adjusting delay discounting procedure showed no changes in impulsivity when atomoxetine was infused into the mPFC or OFC [126]. These data suggest that these two brain regions in isolation cannot account for atomoxetine's observed effects in other studies where impulsivity was shown to decrease upon drug administration.

Using DAT knockout mice another study found that atomoxetine decreased cognitive deficits on an H maze task while having no effect on hyperactivity [127]. This suggests that atomoxetine may improve the cognitive symptoms seen in ADHD without having any action on the dopamine system; however, an effect on the dopamine system may be necessary to decrease symptoms of hyperactivity. Another study was also interested in assessing how atomoxetine may affect the dopamine system as well as how this drug may affect performance on an open field task [128].

In the Moon study SHR were divided in to four atomoxetine treatment groups: control, 1, 5, 0.25 mg/kg/day doses (oral administration). The animals were then assessed in the open field at one, two, and three-week intervals and after the experiment their D2 receptor concentration from the PFC, striatum, and hypothalamus were analyzed using immunohistochemistry. It was found that the 1 mg/kg/day dose significantly decreased open field hyperactivity in the SHR and that D2 receptor concentration decreased in all brain regions in a dose dependent manner. These data suggest that atomoxetine may decrease hyperactivity by normalizing the dopamine system [128]. It is worth noting again that evidence suggests that normalizing the dopamine system may also normalize the glutamate system and vise versa thus glutamate function may play a crucial role here as well [14].

SHR treated with atomoxetine in adolescence self-administer cocaine to a lesser degree compared to those treated with methylphenidate suggesting a decrease in potential drug abuse later in life for those taking this medication [129, 130]. Thus, all things considered, atomoxetine may not only be superior in treating ADHD symptoms compared to traditional treatments but may also be less likely to promote abuse behaviors in the future.

Atomoxetine and ADHD: Evidence from Clinical Trials

Although effect sizes are usually not as large as with stimulants, several clinical trials have been conducted on atomoxetine in the ADHD population with the vast majority showing positive results [131]. For example, a six-month, placebo controlled, double-blind trial showed an improvement in attention scores in adults with ADHD compared to placebo controls [132]. In another double-blind, placebo-controlled study it was found that once-daily atomoxetine improved ADHD symptoms in adolescents according to investigator, parent, and teacher ratings [133]. A meta-analysis conducted on 25 double-blind, placebo controlled studies also showed that atomoxetine decreased several symptoms associated with ADHD such as hyperactivity, impulsivity, and inattention compared to placebo in children and adolescents [134].

An integrated analysis of three Eli Lilly clinical studies showed that that atomoxetine increased emotional control in those with ADHD and that emotional control was also found to correlate to improvements in core ADHD symptoms [135]. Note that it has also been suggested that atomoxetine is equally efficacious regardless of prior stimulant medication treatment although crossover studies are currently occurring to address this question [136].

In a 3-year open label study atomoxetine was found to be more effective in adult, female ADHD patients that also had other emotional dysregulations [137]. Further, an 8-week open-label study in Japanese adults with ADHD showed a significant improvement in ADHD symptoms and a low medication discontinuation rate [138]. In a longer open-label study (12-weeks) atomoxetine was found to be an effective treatment in adults with ADHD that also had comorbid responsive generalized anxiety disorder [139]. A meta-analysis conducted on data from 13 atomoxetine studies (placebo controlled and open-label) further concluded that this drug decreases ADHD symptoms in adolescents with little to no drug tolerance or adverse events after two years of use [140]. Overall there seems to be ample evidence that atomoxetine is safe and effective in children, adolescence, and adults with ADHD.

Memantine: Basic Pharmacology

Memantine is an uncompetitive NMDA antagonist with low to moderate affinity for the NMDA receptor; initially this drug was developed for the treatment of Alzheimer's disease [141]. Pharmacokinetically, memantine is absorbed readily from the gut reaching maximum plasma concentration in 3-8 hours with a half-life of approximately 60-80 hours [110]. Most of the drug is metabolized by the kidneys with very little of the drug being metabolized in the liver by cytochrome p450 enzymes [142]. Memantine appears to have little side effects with the most commonly reported being dizziness, constipation, headache, hypertension, and somnolence [143]. Considering increased glutamate levels are suggested in those with ADHD, it is theorized that memantine's ability to decrease glutamate signaling is a possible reason for its therapeutic value in ADHD [14].

Memantine and ADHD: Evidence from Animal Models

There seems to be a paucity of data assessing memantine's effects on impulsivity in animal models; however, the studies found present mixed results. For example, one study showed that the time spent in the central area on a locomotor task increased in both SHR and WKY controls after a high dose of memantine (32 mg/kg) suggesting an increase in impulsivity [144]. However, at low memantine doses (5.6 mg/kg) the SHR group spent a smaller amount of time in the central area suggesting a decrease in impulsivity [144]. In this same study memantine seemed to have little to no effect on SHR and WKY rats in delay discounting suggesting that this drug did little to impulsivity [144].

In another study memantine seem to increase impulsive choice in low impulsive rats [145]. However, the fact that these authors used a median split in order to divide their rats into low and high impulsive groups may confounded their results; thus, this should be considered when interpreting their data. All things considered, animal models assessing the effects of memantine on impulsivity are mixed; however, clinical trials seem to present more consistent results.

Memantine and ADHD: Evidence from Clinical Trials

Clinical trials have assessed the efficacy of memantine for the treatment of ADHD in children, adolescence, and adults [111, 146]. An eight-week, open-label trial found that a dose of 10 mg/day and 20 mg/day showed dose dependent benefits in both inattention and hyperactivity/impulsivity in adolescents with ADHD-C [111]. Another open-label study showed that memantine titrated to a dose of 10 mg twice a day meliorated ADHD symptoms in adults [146].

In a randomized, double-blind, placebo-controlled clinical trial memantine in conjunction with stimulant medications were shown to improve behavior ratings on ADHD inventories in adult ADHD [147]. To the authors' knowledge, these are the few clinical trials that have assessed memantine in the ADHD population directly. However, there have been other trials assessing memantine in impulsive-like disorders such as gambling, drug addiction, and kleptomania [147, 148].

In a study by Grant et al. (2010) 28 patients diagnosed with pathological gambling received between 10-30 mg/day of memantine for 10 weeks. After study completion participants reported a decrease in compulsive and impulsive behaviors as well as improved cognitive flexibility [149]. In other studies assessing those with substance abuse disorders it was found that memantine reduces cue-induced craving in alcoholics as well as decreases withdrawal symptoms associated with opiate abstinence [150, 151]. Another study showed that a dose of 10 mg/day titrated to 30 mg/day of memantine decreased impulsive stealing behavior in those with kleptomania [148]. Thus, memantine may disrupt impulsive like behaviors associated with ADHD, gambling, and addiction. Further, memantine's action on the glutamate system is likely responsible for this therapeutic effect.

FUTURE PERSPECIVE FOR GLUTAMATE MODULATING DRUGS FOR THE TREATEMENT OF ADHD

The data reviewed in this chapter presents ample evidence of glutamatergic dysfunction in ADHD. Further, data supports that using pharmaceuticals that target the glutamate system (i.e., memantine and atomoxetine) may stabilize glutamatergic signaling and meliorate the behavioral issues associated with ADHD in both animal models and humans. Another interesting thought presented in this chapter is the relatively new idea that the neuro-transmitter imbalances seen in ADHD could be due to issues with how the brain obtains energy. Thus, future pharmaceuticals for the treatment of ADHD targeting the glutamate system and systems that can increase energy flow to the brain could be beneficial.

As previously mentioned, when glutamate is released from synapses and taken up by astrocytes lactate is shuttled from astrocytes to fuel neuronal firing [152]. Further, stimulation of β-adrenergic receptors on astrocytes stimulates glycogenolysis causing an increase in astrocytic glucose that is transformed to lactate and shuttled to neurons [153]. Also, stimulation of α2A-aderendergic receptors on astrocytes may increase astrocytic glycolysis in the long-term thus increasing their energy storage [154]. This becomes interesting considering that a current treatment for ADHD (i.e., atomoxetine) works on both the norepinephrine, dopamine, and glutamate pathways and that a current hypothesis for ADHD is dopamine/norepinephrine hypo-function that in turn may cause glutamatergic hyper-function [14].

Considering the above, it is easy to imagine a situation where lack of brain energy eventually causes catecholamine hypo-function because not enough energy is available to sustain normal neuronal firing. This catecholamine hypo-function then promotes an increase in glutamatergic firing which then puts more energy stress on an already vulnerable system. This could create a feed-forward pathological situation that may lead to the symptoms seen in ADHD as well as other psychiatric illnesses. Thus, the development of drugs which act similar to atomoxetine both to decrease glutamate signaling (i.e., glutamate antagonists) and increase catecholamine signaling (i.e., agonists or uptake inhibitors) may help to increase the energy supply to the brain and directly act on the neurotransmitter systems dysregulated; both effects should normalize brain physiology and meliorate the symptoms of ADHD.

As proposed by Todd and Botteron in 2001, if we consider ADHD to be primarily an issue with how the brain gets energy and that this energy problem causes neurotransmitter dysregulations then a simplistic possibility is that diet and exercise may also help to alleviate symptoms seen in ADHD. The current research from animal models suggests that the data on modulating diet to improve ADHD symptoms is mixed [155, 156]. However, the evidence from human studies seems promising [157]. Although clinical anecdotes from patients with ADHD attest to the importance of exercise in reducing symptoms, evidence from animal models and human studies on exercise and ADHD are mixed but seem promising [158-161]. Thus, perhaps in the future a standard of care for ADHD will not only be using pharmaceuticals that work to increase brain energy but also by prescribing a certain diet and exercise regimen in order to normalize glutamate and catecholamine neurotransmission in the brain.

Conclusion

In this chapter we have reviewed current, relevant data from animal models and humans on the glutamate system and how this system may be dysregulated in ADHD. We have also alluded to how other neurotransmitter systems such as the dopamine and norepinephrine systems may interact with the glutamate system to produce ADHD symptoms. Along with these ideas, the thought that an overarching cause of all of the neurotransmitter dysregulations seen in ADHD may be issues with brain energetics. Current pharmacotherapies that work on the glutamate system were discussed, as were thoughts for future treatments in ADHD that work on the glutamate system either directly or indirectly. It is the hope of the authors that the data reviewed in this book chapter as well as the novel ideas discussed will be helpful to researchers currently working on treatments for ADHD and will encourage further research in this field.

Acknowledgments

The authors would like to thank Hayden Scott and Luke Sparks for their help in obtaining some of the literature reviewed in this book chapter.

References

[1] Martins, A. C. G., Silva de Souza, N., Machado, D. C. D., Orsini, V. H. B. M. & Barbosa, A. D. M. (2014). Historical and diagnostic trajectory

[2] Granet, D. B., Gomi, C. F., Ventura, R. & Miller-Scholte, A. (2005). The relationship between convergence insuffency and ADHD. *Strabismus.*, *13*, 163-8.

[3] Morgan, A. E., Hynd, G. W., Riccio, C. A. & Hall, J. (1996). Validity of DSM-IV ADHD predominantly inattentive and combined types: realationship to previous DSM diagnoses/subtype differences. *J. Am. Aca. Child. Adolesc. Psychiatry.*, *35*, 325-33.

[4] American Psychiatric Association (2013). Diagnostic and statstical manual for mental disorders: DSM-5. American Psychiatric Association., 991.

[5] Mash, E. J. & Barkley, R. A. (2003). Child psychopathology. Guilford Press, 802.

[6] Genro, J. P., Roman. T., Rohde, L. A. & Hutz, M. H. (2012). The Brazilian contribution to attention-deficit/hyperactivity disorder molecular genetics in children and adolescents. *Genet. Mol. Bio.*, *35*, 932-8.

[7] Biederman, J. (2005). Attention-deficit/hyperactivity disorder: a selective overview. *Biol. Psychiatr.*, *57*, 1215-20.

[8] Shimizu, V. T. & Miranda, M. C. (2012). Sensory processing in children with ADHD: a revision of the literature. *Rev. Psicopedag.*, *29*, 256-68.

[9] Zhang, L., Chang, S. & Li, Z. (2012). ADHD gene: a genetic database for attention deficit hyperactivity disorder. *Nucleic Acids Res.*, *40*, D1003-9.

[10] Roman, T., Rohde, L. A. & Hutz, M. H. (2002). Susceptibility genes in attention/deficit hyperactivity disorder. *Rev. Bras. Psiquiatr.*, *24*, 196-201.

[11] Kirley, A., Hawi, Z., Daly, G., McCarron, M., Mullins, C., Millar, N., Waldman, I. & Fitzgerald, M. Gill, M. (2002). Dopaminergic system genes in ADHD: toward a biological hypothesis. *Neuropsychopharmacol.*, *27*, 607-19.

[12] Faraone, S. V., Doyle, A. E., Mick, E. & Biederman, J. (2001). Meta-analysis of the association between the 7-repeat allele of the dopamine D4 receptor gene and attention deficit hyperactivity disorder. *Am. Psychiatr.*, *158*, 1052-7.

[13] Miller, E. M., Pomerleau, F., Huettl, P., Gerhardt, G. A. & Glaser, P. E. (2014). Aberrant glutamate signaling in the prefrontal cortex and striatum of the spontaneously hypertensive rat model of attention-deficit/hyperactivity disorder. *Psychopharmacology.*, *231*, 3019-29.

[14] Miller, E. M., Thomas, T. C., Gerhardt, G. A. & Glaser, P. E. A. (2013). Dopamine and glutamate interactions in ADHD: implications for the future neuropharmacology of ADHD. InTech., 1-30.

[15] Reif, A., Kiive, E., Kurrikoff, T., Paaver, M., Herterich, S., Konstabel, K., Tulviste, T., Lesch, K. P. & Harro, J. (2009). Influence of functional variant of neuronal nitric oxide synthase on impulsive behaviors in humans. *Arch. Gen. Psychiatry.*, 2009, *66*, 41-50.

[16] Hoogman, M., Aarts, E., Zwiers, M., Slaats-Willemse, D., Naber, M., Onnink, M., Cools, R., Kan, C., Buitelaar, J. & Franke, B. (2011). Nitric oxide synthase genotype modulation of impulsivity and ventral striatal activity in adult ADHD patients and healthy comparison subjects. *Am. J. Psychiatry.*, *168*, 1099-106.

[17] Rohde, L. A., Miguel Filho, E. C., Benetti, L., Gallois, C. & Kieling, C. (2004). Attention-deficit/hyperactivity disorder in childhood and adolescence: clinical and therapeutic aspects. *Rev. Psiq. Cl.*, *31*, 124-31.

[18] Yoshimasu, K., Barbaresi, W. J., Colligan, R. C., Voigt, R. G., Killian, J. M., Weaver, A. L. & Katusic S. K. (2012). Childhood ADHD is strongly associated with a broad range of psychiatric disorders during adolescence: a population-based birth cohort study. *J. Child Psychol. Psychiatry.*, *53*, 1036-43.

[19] Rafolovich, A. (2001). The conceptual history of attention deficit hyperactivity disorder: idiocy, imbecility, encephalitis, and the child deviant 1877-1929. *Deviant Behav.*, *22*, 93-115.

[20] Chen, M. H., Chen, Y. S., Hsu, J. W., Huang, K. L., Li, C. T., Lin, W. C., Chang, W. H., Chen, T. J., Pan, T. L., Su, T. P. & Bai, Y. M. (2015). Comorbidity of ADHD and subsequent bipolar disorder among adolescents and young adults with major depression: a nationwide longitudinal study. *Bipolar Disord.*, *17*, 315-22.

[21] Carrey, N., MacMaster, F. P., Sparkes, S. J., Kahn, S. C. & Kusumaker, V. (2002). Glutamatergic changes with treatment in attention deficit hyperactivity disorder: a preliminary case series. *J. Child Adolesc. Psychopharmacol.*, *12*, 331-6.

[22] Moore, C. M., Biederman, J., Wozniak, J., Mick, E., Aleardi, M., Wardrop, M., Dougherty, M., Harpold, T., Hammerness, P., Randall, E. & Renshaw, P. F. (2006). Differences in brain chemistry in children and adolescents with attention deficit hyperactivity disorder with and without comorbid bipolar disorder: a proton magnetic resonance spectroscopy study. *Am. J. Psychiatry.*, *163*, 316-8.

[23] Moore, C. M., Frazier, J. A., Glod, C. A., Breeze, J. L., Dieterich, M., Finn, C. T., Frederick, B. & Renshaw, P. F. (2007). Glutamine and glutamate levels in children and adolescents with bipolar disorder: a 4.0-T proton magnetic resonance spectroscopy study of the anterior cingulate cortex. *J. Am. Acad. Child Adolesc. Psychiatry.*, *46*, 524-34.

[24] Miller, E. M. (2014). Dopamine and glutamate dysfunction in a rodent model of attention-deficit/hyperactivity disorder: implications for future neuropharmacology. Theses and Dissertations--Anatomy and Neurobiology.

[25] Danbolt, N. C. (2001). Glutamate uptake. *Prog. Neurobiol.*, *65*, 1-105.

[26] Tapiero, H., Mathe, H., Couvreur, P. & Tew, K. D. (2002). Glutamine and glutamate. *Biomed. and Pharmacother.*, *56*, 446-57.

[27] Anderson, C. M. & Swanson, R. A. (2000). Astrocyte glutamate transport: review of properties, regulation, and physiological functions. *Glia.*, *32*, 1-14.

[28] Daikhin, Y. & Yudkoff, M. (2000). Compartmentalization of brain glutamate metabolism in neurons and glia. *Journal Nutr.*, *130*, 1026S-31S.

[29] Fonnum, F., Fykse, E. M. & Roseth, S. (1998). Chapter 7: Uptake of glutamate into synaptic vesicles. *Prog. Brain Res.*, *116*, 87-101.

[30] Meldrum, B. S. (2000). Glutamate as a neurotransmitter in the brain: review of physiology and pathology. *Journal of Nutr.*, 1007S-15S.

[31] Turner, T. J. (1998). Chapter 1: Calcium channels coupled to glutamate release. *Prog. Brain Res.*, *116*, 3-14.

[32] Attwell, D. (2000). Brain uptake of glutamate: food for thought. *Journal Nutr.*, *130*, 1023S-25S.

[33] Iversen, L. L., Iversen, S. D., Bloom, F. E. & Roth, R. H. (2009). Introduction to neuropsychopharmacology. New York: Oxford University Press.

[34] Schoepp, D. D., Jane, D. E. & Monn, J. A. (1998). Pharmacological agents acting at subtypes of metabotropic glutamate receptors. *Neuropharmacology.*, *38*, 1431-76.

[35] Danbolt, N. C., Chaudhry F. A., Dehnes, Y., Lehre, K. P., Levy, K. P. Ullensvang, K. & Storm-Mathisen, J. (1998). Chapter 3: Properties and localization of glutamate transporters. *Prog. Brain Res.*, *116*, 23-43.

[36] Seal, R. P. & Amara, S. G. (1999). Excitatory amino acid transporters: a family in flux. *Annu. Rev. Pharmacol.*, *39*, 431-56.

[37] Schultz, W. (2001). Reward signaling by dopamine neurons. *Neuroscientist.*, *7*, 293-302.

[38] Martinez-Fong, D., Rosales, M. G., Góngora-Alfaro, J., Hernández, S. & Aceves, G. (1992). NMDA receptor mediates dopamine release in the striatum of unanesthetized rats as measured by brain microdialysis. *Brain Res.*, *595*, 309-15.

[39] Warton, F. L., Howells, F. M. & Russell, V. A. (2009). Increased glutamate-stimulated release of dopamine in substantia nigra of a rat model for attention-deficit/hyperactivity disorder—lack of effect of methylphenidate. *Metab. Brain Dis.*, *24*, 599-613.

[40] Kotecha, S. A., Oak, J. N., Jackson, M. F., Perez, Y., Orser, B. A., Van Tol H. H. M. & MacDonald, J. F. (2002). A D2 class dopamine receptor transactivates a receptor tyrosine kinase to inhibit NMDA receptor transmission. *Neuron.*, *35*, 11-22.

[41] Yuen, E. Y., Liu, W., Kafri, T., van Praag, H. & Yan, Z. (2010). Regulation of AMPA receptor channels and synaptic plasticity by cofilin phosphatase Slingshot in cortical neurons. *J. Physiol.*, *588*, 2361-71.

[42] Castellanos, F. X. & Tannock, R. (2002). Neuroscience of attention-deficit/hyperactivity disorder: the search for endophenotypes. *Nat. Rev. Neurosci.*, *3*, 617-28.

[43] Tannock, R., Ickowicz, A. & Schachar, R. (1995). Differential effects of methylphenidate on working memory in ADHD children with and without comorbid anxiety. *J. Am. Acad. Child Adolesc. Psychiatry.*, *34*, 886-96.

[44] Hammerness, P., Biederman, J., Petty, C., Henin, A. & Moore, C. M. (2012). Brain biochemical effects of methylphenidate treatment using proton magnetic spectroscopy in youth with attention-deficit hyperactivity disorder: a controlled pilot study. *CNS Neurosci. Ther.*, *18*, 34-40.

[45] Sagvolden, T. & Johansen, E. B. (2011). Rat models of ADHD. *Curr. Top. Behav. Neurosci.*, *9*, 301-15.

[46] Russell, V. A. (2001). Chapter 9: Overview of animal models of attention deficit hyperactivity disorder (ADHD). *Curr. Protoc. Neurosci.*, *54*, 1-25.

[47] Sagvolden, T. (2000). Behavioral validation of the spontaneously hypertensive rat (SHR) as an animal model of attention-deficit/hyperactivity disorder (AD/HD). *Neurosci. Biobehav. Rev.*, *24*, 31-39.

[48] Sagvolden, T., Hendley, E. D. & Knardahl, S. (1992). Behavior of hypertensive and hyperactive rat strains: hyperactivity is not unitarily determined. *Physiol. Behav.*, *52*, 49-57.

[49] Knardahl, S. & Sagvolden, T. (1979). Open-field behavior of spontaneously hypertensive rats. *Behav. Neural Biol.*, *27*, 187-200.

[50] Alsop, B. (2007). Problems with spontaneously hypertensive rats (SHR) as a model of attention-deficit/hyperactivity disorder (AD/HD). *J. Neurosci. Methods.*, *162*, 42-48.

[51] Sagvolden, T., Russell, V. A., Aase, H., Johansen, E. B. & Farshbaf, M. (2005). Rodent models of attention-deficit/hyperactivity disorder. *Biol. Psychiatry.*, *57*, 1239-47.

[52] Sagvolden, T., Dasbanerjee, T., Zhang-James, Y., Middleton, F. & Faraone, S. (2008). Behavioral and genetic evidence for a novel animal model of Attention-Deficit/Hyperactivity Disorder predominantly inattentive subtype. *Behav. Brain Funct.*, *4*, 56.

[53] Sagvolden, T., Johansen, E. B., Woien, G., Walaas, S. I., Storm-Mathisen, J., Bergersen, L. H., Hvalby, O., Jensen, V., Aase, H., Russell, V. A., Killeen, P. R., Dasbanerjee, T., Middleton, F. A. & Faraone, S. V. (2009). The spontaneously hypertensive rat model of ADHD--the importance of selecting the appropriate reference strain. *Neuropharmacology.*, *57*, 619-26.

[54] Granziera, C., Nouchine, H., Shahar, A., Margitta, S., Reto, M. & Gunnar, K. (2011). In-vivo magnetic resonance imaging of the structural core of the Papez circuit in humans. *Neuroreport.*, *22*, 227-31.

[55] Adey, W. R. & Meyer, M. (1952). An experimental study of hippocampal afferent pathways from prefrontal and cingulate areas in the monkey. *J. Anat.*, *86*, 58-74.

[56] Dalley, J. W., Cardinal, R. N. & Robbins, T. W. (2004). Prefrontal executive and cognitive functions in rodents: neural and neurochemical substrates. *Neurosci. Biobehav. Rev.*, *28*, 771-84.

[57] Krusch, D. A., Klorman, R., Brumaghim, J. T., Fitzpatrick, P. A., Borgstedt, A. D. & Strauss, J. (1996). Methylphenidate slows reactions of children with attention deficit disorder during and after an error. J. Abnorm. *Child Psychol.*, *24*, 633-50.

[58] Mehta, M. A., Goodyer, I. M. & Sahakian, B. J. (2004). Methylphenidate improves working memory and set-shifting in AD/HD: relationships to baseline memory capacity. *J. Child Psychol. Psychiatry.*, *45*, 293-305.

[59] Klimkeit, E. I., Mattingley, J. B., Sheppard, D. M., Lee, P. & Bradshaw, J. L. (2005). Motor preparation, motor execution, attention, and executive functions in attention deficit/hyperactivity disorder (ADHD). *Child Neuropsychol.*, *11*, 153-73.

[60] Dunnett, S. B. & Lelos, M. (2010). Behavioral analysis of motor and non-motor symptoms in rodent models of Parkinson's disease. *Prog. Brain Res.*, *184*, 35-51.

[61] Salamone, J. D. & Correa, M. (2012). The mysterious motivational functions of mesolimbic dopamine. *Neuron.*, *76*, 470-85.

[62] Lehohla, M., Kellaway, L. & Russell, V. A. (2004). NMDA receptor function in the prefrontal cortex of a rat model for attention-deficit hyperactivity disorder. *Metab. Brain Dis.*, *19*, 35-42.

[63] Russell, V. A. (2001). Increased AMPA receptor function in slices containing the prefrontal cortex of spontaneously hypertensive rats. *Metab. Brain Dis.*, *16*, 143-49.

[64] Gainetdinov, R. R. (1999). Role of serotonin in the paradoxical calming effect of psychostimulants on hyperactivity. *Science.*, *283*, 397-401.

[65] Gainetdinov, R. R. (2000). An animal model of attention deficit hyperactivity disorder. *Mol. Med. Today.*, *6*, 43-44.

[66] Masuo, Y., Ishido, M., Morita, M. & Oka, S. (2002). Effects of neonatal 6-hydroxydopamine lesion on the gene expression profile in young adult rats. *Neurosci. Lett.*, *335*, 124-28.

[67] Halberstadt, A. L., Lehmann-Masten, V. D., Geyer, M. A. & Powell, S. B. (2011). Interactive effects of mGlu5 and 5-HT2A receptors on locomotor activity in mice. *Psychopharmacology.*, *215*, 81-92.

[68] Benn, A. & Robinson, E. S. (2014). Investigating glutamatergic mechanism in attention and impulse control using rats in a modified 5-choice serial reaction time task. *PLoS One.*, *9*, e115374.

[69] Li, D. P., Pachuau, J., Lee, H. A. & Pan, H. L. (2014). mGluR5 upregulation increases excitability of hypothalamic presympathetic neurons through NMDA receptor trafficking in spontaneously hypertensive rats. *J. Neurosci.*, *34*, 4309-17.

[70] Yabuki, Y., Shioda, N., Maeda, T., Hiraide, S., Togashi, H. & Fukunaga, K. (2014). Aberrant CaMKII activity in the medial prefrontal cortex is associated with cognitive dysfunction in ADHD model rats. *Brain Res.*, *1557*, 90-100.

[71] Yates, J. R., Batten, S. R., Bardo, M. T. & Beckmann, J. S. (2015). Role of ionotropic glutamate receptors in delay and probability discounting in the rat. *Psychopharmacology.*, *232*, 1187-96.

[72] Perry, J. L. & Carroll, M. E. (2008). The role of impulsive behavior in drug abuse. *Psychopharmacology.*, *200*, 1-26.

[73] Silva, J. P., Lelianova, V. G., Ermolyuk, Y. S., Vysokov, N., Hitchen, P. G., Berninghausen, O., Rahman, M. A., Zangrandi, A., Fidalgo, S., Tonevitsky, A. G., Dell, A., Volynski, K. E. & Ushkaryov, Y. A. (2011). Latrophilin 1 and its endogenous ligand Lasso/teneurin-2 form a high affinity transsynaptic receptor pair with signaling capabilities. *Prco. Natl. Acad. Sci.*, *108*, 12113-18.

[74] Lange, M., Norton, W., Coolen, M., Chaminade, M., Merker, S., Proft, F., Schmitt, A., Vernier, P., Lesch, K. P. & Bally-Cuif, L. (2012). The ADHD-susceptibility gene lphn3.1 modulates dopaminergic neuron formation and locomotor activity during zebrafish development. *Mol. Psychiatry.*, *17*, 946-54.

[75] O'Sullivan, M. L., de Wit, J., Savas, J. N., Comoletti, D., Otto-Hitt, S., Yates, J. R. & Ghosh, A. (2012). FLRT proteins are endogenous latrophilin ligands and regulate excitatory synapse development. *Neuron.*, *73*, 903-10.

[76] Yamagishi, S., Hampel, F., Hata, K., Del Toro, D., Schwark, M., Kvachnina, E., Bastmeyer, M., Yamashita, T., Tarabykin, V., Klein, R. & Egea, J. (2011). FLRT2 and FLRT3 act as repulsive guidance cues for Unc5-positive neurons. *EMBO J.*, *30*, 2920-33.

[77] Lionel, A. C., Crosbie, J., Barbosa, N., Goodale, T., Thiruvahindrapuram, B., Rickaby, J., Gazzellone, M., Carson, A. R., Howe, J. L., Wang, Z., Wei, J., Stewart, A. F., Roberts, R., McPherson, R., Fiebig, A., Franke, A., Schreiber, S., Zwaigenbaum, L., Fernandez, B. A., Roberts, W., Arnold, P. D., Szatmari, P., Marshall, C. R., Schachar, R. & Scherer, S. W. (2011). Rare copy number variation discovery and cross-disorder comparisons identify risk genes for ADHD. *Sci. Transl. Med.*, *3*, 1-11.

[78] Durand, C. M., Betancur, C., Boeckers, T. M., Bockmann, J., Chaste, P., Fauchereau, F., Nygren, G., Rastam, M., Gillberg, I. C., Anckarsater, H., Sponheim, E., Goubran-Botros, H., Delorme, R., Chabane, N., Mouren-Simeoni, M. C., de Mas, P., Bieth, E., Roge, B., Heron, D., Burglen, L., Gillberg, C., Leboyer, M. & Bourgeron, T. (2007). Mutations in the gene encoding the synaptic scaffolding protein SHANK3 are associated with autism spectrum disorders. *Nat. Genet.*, *39*, 25-7.

[79] Lesch, K. P., Timmesfeld, N., Renner, T. J., Halperin, R., Roser, C., Nguyen, T. T., Craig, D. W., Romanos, J., Heine, M., Meyer, J., Freitag, C., Warnke, A., Romanos, M., Schafer, H., Walitza, S., Reif, A., Stephan, D. A. & Jacob, C. (2008). Molecular genetics of adult ADHD: converging evidence from genome-wide association and extended pedigree linkage studies. *J. Neural Transm.*, *115*, 1573-85.

[80] Reif, A., Herterich, S., Strobel, A., Ehlis, A. C., Saur, D., Jacob, C. P., Wienker, T., Topner, T., Fritzen, S., Walter, U., Schmitt, A., Fallgatter, A. J. & Lesch, K. P. (2006).

A neuronal nitric oxide synthase (NOS-I) haplotype associated with schizophrenia modifies prefrontal cortex function. *Mol. Psychiatry.*, *11*, 286-300.

[81] Dorval, K. M., Wigg, K. G., Crosbie, J., Tannock, R., Kennedy, J. L., Ickowicz, A., Pathare, T., Malone, M., Schachar, R. & Barr, C. L. (2007). Association of the glutamate receptor subunit gene GRIN2B with attention-deficit/hyperactivity disorder. *Genes Brain Behav.*, *6*, 444-52.

[82] Turic, D., Williams, H., Langley, K., Owen, M., Thapar, A. & O'Donovan, M. C. (2005). A family based study of catechol-O-methyltransferase (COMT) and attention deficit hyperactivity disorder (ADHD). *Am. J. Med. Genet. B. Neuropsychiatr. Genet.*, *133B*, 64-7.

[83] Elia, J., Glessner, J. T., Wang, K., Takahashi, N., Shtir, C. J., Hadley, D., Sleiman, P. M., Zhang, H., Kim, C. E., Robison, R., Lyon, G. J., Flory, J. H., Bradfield, J. P., Imielinski, M., Hou, C., Frackelton, E. C., Chiavacci, R. M., Sakurai, T., Rabin, C., Middleton, F. A., Thomas, K. A., Garris, M., Mentch, F., Freitag, C. M., Steinhausen, H. C., Todorov, A. A., Reif, A., Rothenberger, A., Franke, B., Mick, E. O., Roeyers, H., Buitelaar, J., Lesch, K. P., Banaschewski, T., Ebstein, R. P., Mulas, F., Oades, R. D., Sergeant, J., Sonuga-Barke, E., Renner, T. J., Romanos, M., Romanos, J., Warnke, A., Walitza, S., Meyer, J., Palmason, H., Seitz, C., Loo, S. K., Smalley, S. L., Biederman, J., Kent, L., Asherson, P., Anney, R. J., Gaynor, J. W., Shaw, P., Devoto, M., White, P. S., Grant, S. F., Buxbaum, J. D., Rapoport, J. L., Williams, N. M., Nelson, S. F., Faraone, S. V. & Hakonarson, H. (2012). Genome-wide copy number variation study associates metabotropic glutamate receptor gene networks with attention deficit hyperactivity disorder. *Nat. Genet.*, *44*, 78-84.

[84] Magistretti, P. J., Pellerin, L., Rothman, D. L. & Shulman, R. G. (1999). Energy on demand. *Science.*, *283*, 496-97.

[85] Fonseca-Azevedo, K. & Herculano-Houzel, S. (2012). Metabolic constraint imposes tradeoff between body size and number of brain neurons in human evolution. *PNAS.*, *109*, 18571-76.

[86] Attwell, D. & Laughlin, S. B. (2001). An energy budget for signaling in the grey matter of the brain. *J. Cerebr. Blood F. Met.*, *21*, 1133-45.

[87] Almeida, A., Almeida, J., Bolanos, J. P. & Moncada, S. (2001). Different responses of astrocytes and neurons to nitric oxide: the role of glycolytically generated ATP in astrocyte protection. *Proc. Natl. Acad. Sci.*, *98*, 15294-9.

[88] Barros, L. F. (2013). Metabolic signaling by lactate in the brain. *Trends Neurosci.*, *36*, 396-404.

[89] Magistretti, P. J. (2006). Neuron-glia metabolic coupling and plasticity. *J. Exp. Biol.*, *209*, 2304-11.

[90] Newman, L. A., Korol, D. L. & Gold, P. E. (2011). Lactate produced by glycogenolysis in astrocytes regulates memory processing. *PLoS One.*, *6*, 1-11.

[91] Suzuki, A., Stern, S. A., Bozdagi, O., Huntley, G. W., Walker, R. H., Magistretti, P. J. & Alberini, C. M. (2011). Astrocyte-neuron lactate transport is required for long-term memory formation. *Cell.*, *144*, 810-23.

[92] Yang, J., Ruchti, E., Petit, J. M., Jourdain, P., Grenningloh, G., Allaman, I. & Magistretti, P. J. (2014). Lactate promotes plasticity gene expression by potentiating NMDA signaling in neurons. *Proc. Natl. Acad. Sci.*, *111*, 12228-33.

[93] Todd, R. D. & Botteron, K. N. (2001). Is attention-deficit/hyperactivity disorder and energy deficiency syndrome? *Biol. Psychiatry.*, *50*, 151-58.

[94] Killeen, P. R., Russell, V. A. & Sergeant, J. A. (2013). A behavioral neuroenergetics theory of ADHD. *Neurosci. Biobehav. Rev.*, *37*, 625-57.

[95] Cortese, S., Kelly, C., Chabernaud, C., Proal, E., Di Martino, A., Milham, M. P. & Castellanos, F. X. (2012). Toward systems neuroscience of ADHD: a meta-analysis of 55 fMRI studies. *Am. J. Psychiatry.*, *169*, 1038-55.

[96] Fair, D. A., Posner, J., Nagel, B. J., Bathula, D., Dias, T. G., Mills, K. L., Blythe, M. S., Giwa, A., Schmitt, C. F. & Nigg, J. T. (2010). Atypical default network connectivity in youth with attention-deficit/hyperactivity disorder. *Biol. Psychiatry.*, *68*, 1084-91.

[97] Nagel, B. J., Bathula, D., Herting, M., Schmitt, C., Kroenke, C. D., Fair, D. & Nigg, J. T. (2011). Altered white matter microstructure in children with attention-deficit/hyperactivity disorder. *J. Am. Acad. Child Adolesc. Psychiatry.*, *50*, 283-92.

[98] Koos, T. & Tepper, J. M. (1999). Inhibitor control of neostriatal projections to neurons by GABAergic interneurons. *Nature.*, *2*, 467-72.

[99] Carter, B. C. & Bean, B. P. (2009). Sodium entry during action potentials of mammalian neurons: incomplete inactivation and reduced metabolic efficiency in fast-spiking neurons. *Neuron.*, *64*, 898-909.

[100] Ariyannur, P. S., Moffett, J. R., Manickam, P., Pattabiraman, N., Arun, P., Nitta, A., Nabeshima, T., Madhavarao, C. N. & Namboodiri, A. M. (2010). Methamphetamine-induced neuronal protein NAT8L is the NAA biosynthetic enzyme: implications for specialized acetyl coenzyme A metabolism in the CNS. *Brain Res.*, *1335*, 1-13.

[101] Yang, P., Wu, M. T., Dung, S. S. & Ko, C. W. (2010). Short-TE proton magnetic resonance spectroscopy investigation in adolescents with attention-deficit hyperactivity disorder. *Psychiatry Res.*, *181*, 199-203.

[102] Perlov, E., Philipsen, A., Matthies, S., Drieling, T., Maier, S., Bubl, E., Hesslinger, B., Buechert, M., Henning, J., Ebert, D. & Tebartz Van Elst, L. (2009). Spectroscopic findings in attention-deficit/hyperactivity disorder: review and meta analysis. *World J. Biol. Pscychia.*, *10*, 355-65.

[103] Kronenberg, G., Ende, G., Alm, B., Deuschle, M., Heuser, I. & Colla, M. (2008). Increased NAA and reduced choline levels in the anterior cingulum following chronic methylphenidate. A spectroscopic test-retest study in adult ADHD. *Eur. Arch. Psychiatry Clin. Neurosci.*, *258*, 446-50.

[104] Van Meel, C. S., Heslenfeld, D. J., Oosterlaan, J. & Sergeant, J. A. (2007). Adaptive control deficits in attention-deficit/hyperactivity disorder (ADHD): the role of error processing. *Psychiatry Res.*, *151*, 211-20.

[105] Harris, J. J. & Attwell, D. (2012). The energetics of CNS white matter. *J. Neurosci.*, *32*, 356-71.

[106] Zhu, X. H., Qiao, H., Du, F., Xiong, Q., Liu, X., Zhang, X., Ugurbil, K. & Chen, W. (2012). Quantitative imaging of energy expenditure in human brain. *Neuroimage.*, *60*, 2107-17.

[107] Jolivet, R., Allaman, I., Pellerin, L., Magistretti, P. J. & Weber, B. (2010). Comment on recent modeling studies of astrocyte-neuron metabolic interactions. *J. Cereb. Blood Flow Metab.*, *30*, 1982-6.

[108] Pancani, T., Anderson, K. L., Porter, N. M. & Thibault, O. (2011). Imaging of a glucose analog, calcium and NADH in neurons and astrocytes: dynamic responses to depolarization and sensitivity to pioglitazone. *Cell Calcium.*, *50*, 548-58.

[109] Carrey, N. J., MacMaster, F. P., Gaudet, L. & Schmidt, M. H. (2007). Striatal creatine and glutamate/glutamine in attention-deficit/hyperactivity disorder. *J. Child Adolesc. Psychopharmacol.*, *17*, 11-7.

[110] Kavirajan, H. (2009). Memantine: a comprehensive review of safety and efficacy. *Expert Opin. Drug Saf.*, *8*, 89-109.

[111] Findling, R. L., McNamara, N. K., Stansbrey, R. J., Maxhimer, R., Periclou, A., Mann, A. & Graham, S. M. (2007). A pilot evaluation of the safety, tolerability, pharmacokinetics, and effectiveness of memantine in pediatric patients with attention-deficit/hyperactivity disorder combined type. *J. Child. Adolesc. Psychopharmacol.*, *17*, 19-33.

[112] Jensen, L. S., Pagsberg, A. K. & Dalhoff, K. P. (2015). Difference in abuse potential of ADHD drugs measured by contrasting poison centre data and therapeutic use data. *Clin. Toxicol.*, *53*, 210-214.

[113] Wernicke, J. F. & Kratochvil, C. J. (2002). Safety profile of atomoxetine in the treatment of children and adolescents with ADHD. *J. Clin. Psychiatry.*, *6*, 50-55.

[114] Michelson, D., Read, H. A., Ruff, D. D., Witcher, J., Zhang, S. & McCracken, J. (2007). CYP2D6 and clinical response to atomoxetine in children and adolescents with ADHD. *J. Am. Acad. Child Adolesc. Psychiatry.*, *46*, 242-51.

[115] Ludolph, A. G., Udvardi, P. T., Schaz, U., Henes, C., Adolph, O., Weigt, H. U., Fegert, J. M., Boeckers, T. M. & Fohr, K. J. (2010). Atomoxetine acts as an NMDA receptor blocker in clinically relevant concentrations. *Br. J. Pharmacol.*, *160*, 283-91.

[116] Miceli, M. D. & Gronier, B. (2015). Psychostimulants and atomoxetine alter the electrophysiological activity of prefrontal cortex neurons, interaction with catecholamine and glutamate NMDA receptors. *Psychopharmacology.*, *232*, 2191-205.

[117] Witcher, J. W., Long, A., Smith, B., Sauer, J. M., Heilgenstein, J., Wilens, T., Spencer, T. & Biederman, J. (2003). Atomoxetine pharmacokinetics in children and adolescents with attention deficit hyperactivity disorder. *J. Child Adoles. Psychpharmacol.*, *3*, 53-63.

[118] Sauer, J. M., Ring, B. J. & Witcher, J. W. (2005). Clinical pharmacokinetics of atomoxetine. *Clin. Pharmacokinet.*, *44*, 571-590.

[119] Liu, Y. P., Huang, T. S., Tung, C. S. & Lin, C. C. (2015). Effects of atomoxetine on attention and impulsivity in the five-choice serial reaction time task in rats with lesions of dorsal noradrenergic ascending bundle. *Prog. Neuropsychopharmacol. Biol. Psychiatry.*, *56*, 81-90.

[120] Ansquer, S., Belin-Rauscent, A., Dugast, E., Duran, T., Benatru, I., Mar, A. C., Houeto, J. L. & Belin, D. (2014). Atomoxetine decreases vulnerability to develop compulsivity in high impulsive rats. *Biol. Psychiatry.*, *75*, 825-32.

[121] Dommett, E. J. (2014). Using the five-choice serial reaction time task to examine the effects of atomoxetine and methylphenidate in the male spontaneously hypertensive rat. *Pharmacol. Biochem. Be.*, *124*, 196-203.

[122] Broos, N., Schmaal, L., Wiskerke, J., Kostelijk, L., Lam, T., Stoop, N., Weierink, L., Ham, J., de Geus, E. J., Schoffelmeer, A. N., van den Brink, W., Veltman, D. J., de

Vries, T. J., Pattij, T. & Goudriaan, A. E. (2012). The relationship between impulsive choice and impulsive action: a cross-species translational study. *PLoS One.*, *7*, 1-9.

[123] Bayless, D. W., Perez, M. C. & Daniel, J. M. (2015). Comparison of the validity of the use of the spontaneously hypertensive rat as a model of attention deficit hyperactivity disorder in males and females. *Behav. Brain Res.*, *286*, 85-92.

[124] Shibasaki, Y., Hayata-Takano, A., Hazama, K., Nakazawa, T., Shintani, N., Kasai, A., Nagayasu, K., Hashimoto, R., Tanida, M., Katayama, T., Matsuzaki, S., Yamada, K., Taniike, M., Onaka, Y., Ago, Y., Waschek, J. A., Koves, K., Reglodi, D., Tamas, A., Matsuda, T., Baba, A. & Hashimoto, H. (2015). Atomoxetine reverses locomotor hyperactivity, impaired novel object recognition, and prepulse inhibition impairment in mice lacking pituitary adenylate cyclase-activating polypeptide. *Neuroscience.*, *297*, 95-104.

[125] Montes, D. R., Stopper, C. M. & Floresco, S. B. (2015). Noradrenergic modulation of risk/reward decision making. *Psychopharmacology.*, 1-15.

[126] Yates, J. R., Perry, J. L., Meyer, A. C., Gipson, C. D., Charnigo, R. & Bardo, M. T. (2014). Role of medial prefrontal and orbitofrontal monoamine transporters and receptors in performance in an adjusting delay discounting procedure. *Brain Res.*, *1574*, 26-36.

[127] Del'Guidice, T., Lemasson, M., Etievant, A., Manta, S., Magno, L. A., Escoffier, G., Roman, F. S. & Beaulieu, J. M. (2014). Dissociations between cognitive and motor effects of psychostimulants and atomoxetine in hyperactive DAT-KO mice. *Psychopharmacology.*, *231*, 109-22.

[128] Moon, S. J., Kim, C. J., Lee, Y. J., Hong, M., Han, J. & Bahn, G. H. (2014). Effect of atomoxetine on hyperactivity in an animal model of attention-deficit/hyperactivity disorder (ADHD). *PLoS One.*, *9*, 1-6.

[129] Jordan, C. J., Harvey, R. C., Baskin, B. B., Dwoskin, L. P. & Kantak, K. M. (2014). Cocaine-seeking behavior in a genetic model of attention-deficit/hyperactivity disorder following adolescent methylphenidate or atomoxetine treatments. *Drug Alcohol Depend.*, *140*, 25-32.

[130] Jordan, C. J., Harvey, R. C., Baskin, B. M., Rodriguez, K., Dellamorte, A., Dwoskin, L. P. & Kantak, K. M. (2015). Adolescent atomoxetine, but not methylphenidate, decreases cocaine-seeking behavior in a genetic model of attention-deficit/hyperactivity disorder. *Drug and Alcohol Depend.*, *146*, e118-e201.

[131] Bitter, I., Angyalosi, A. & Czobor, P. (2012). Pharmacological treatment of adult ADHD. *Curr. Opin. Psychiatry.*, *25*, 529-34.

[132] Brown, T. E., Holdnack, J., Saylor, K., Adler, L., Spencer, T., Williams, D. W., Padival, A. K., Schuh, K., Trzepacz, P. T. & Kelsey, D. (2011). Effect of atomoxetine on executive function impairments in adults with ADHD. *J. Attent. Dis.*, *15*, 130-38.

[133] Michelson, D., Read, H. A., Ruff, D. D., Witcher, J., Zhang, S. & McCracken, J. (2007). CYP2D6 and clinical response to atomoxetine in children and adolescents with ADHD. *J. Am. Acad. Child Adolesc. Psychiatry.*, *46*, 242-51.

[134] Schwartz, S. & Correll, C. U. (2014). Efficacy and safety of atomoxetine in children and adolescents with attention-deficit/hyperactivity disorder: results from a comprehensive meta-analysis and metaregression. *J. Am. Acad. Child Adolesc. Psychiatry.*, *253*, 174-87.

[135] Asherson, P., Stes, S., Nilsson, M., Berggren, L., Svanborg, P., Kutzelnigg, A. & Deberdt, W. (2015). The effects of atomoxetine on emotional control in adults with ADHD: An integrated analysis of multicenter studies. *Eur. Psychiatry.*, doi: 10.1016/j.eurpsy.2014.12.002.

[136] Wehmeier, P. M., Dittmann, R. W., Banaschewski, T. & Schacht, A. (2014). Does stimulant pretreatment modify atomoxetine effects on core symptoms of ADHD in children assessed by quantitative measurement technology? *J. Attent. Dis.*, *18*, 105-116.

[137] Marchannt, B. K., Reimherr, F. W., Halls, C., Williams, E. D., Strong, R. E., Kondo, D., Soni, P. & Robison, R. J. (2011). Long-term open-label response to atomoxetine in adult ADHD: influence of sex, emotional dysregulation, and double-blind response to amoxetine. *Atten. Defic. Hyperact. Disord.*, *3*, 237-44.

[138] Takahashi, M., Goto, T., Takita, Y., Chung, S. K., Wang, Y. G. & Shur-Fen, S. (2014). Open-label, dose-titration tolerability study of atomoxetine hydrochloride in Korean, Chinese, and Taiwanese adults with attention-deficit/hyperactivity disorder. *Asia Pac. Psychiatry.*, *6*, 62-70.

[139] Gabriel, A. & Violato, C. (2011). Adjunctive atomoxetine to SSRIs or SNRIs in the treatment of adult ADHD patients with comorbid partially responsive generalized anxiety (GA): an open-label study. *Ann. Ny. Acad. Sci.*, *3*, 319-26.

[140] Wilens, T. E., Newcorn, J. H, Kratochvil, C. J., Gao, H., Thomason C. K., Rogers, A. K., Feldman, P. D. & Levine, L. R. (2006). Long-term atomoxetine treatment in adolescents with attention-deficit/hyperactivity disorder. *J. Pediatr.*, *149*, 112-19.

[141] Maeng S. & Zarate, C. A. (2007). The role of glutamate in mood disorders: results from the ketamine in major depression study and the presumed cellular mechanism Underlying its Antidepressant Effects. *Psychia. Rep.*, *9*, 467-74.

[142] Kornhuber, J., Kennepohl, E.M., Bleich, S., Wiltfang, J., Kraus, T., Reulbach, U. & Meineke, I. (2007). Memantine pharmacotherapy. A naturalistic study using a population pharmacokinetic approach. *Clinical Pharmacokinetics.*, *46*, 599-612.

[143] Sani, G., Serra, G., Kotzalidis, G. D., Romano, S., Tamorri, S. M., Manfredi, G., Caloro, M., Telesforo, C. L., Caltagirone, S. S., Panaccione, I., Simonetti, A., Demontis, F., Serra, G. & Girardi, P. (2012). The role of memantine in the treatment of psychiatric disorders other than the dementias: A review of current preclinical and clinical evidence. *CNS Drugs.*, *26*, 663-90.

[144] Sukhanov, I. M., Zakharova, E. S., Danysz, W. & Bespalov, A. Y. (2004). Effects of NMDA receptor channel blockers, MK-801 and memantine, on locomotor activity and tolerance to delay of reward in Wistar Kyoto and spontaneously hypertensive rats. *Behavioural Pharmacology.*, *15*, 263-71.

[145] Cottone, P., Lemolo, A., Narayan, A. R., Kwak, J., Momaney, D. & Sabino, V. (2013). The uncompetitive NMDA receptor antagonists ketamine and memantine preferentially increase the choice for a small, immediate reward in low-impulsive rats. *Psychopharmacology.*, *226*, 127-38.

[146] Surman, C. B., Hammerness, P. G., Petty, C., Spencer, T., Doyle, R., Napolean, S., Chu, N., Yorks, D. & Biederman, J. (2013). A pilot open label prospective study of memantine monotherapy in adults with ADHD. *World J. Biol. Psychiatry.*, *14*, 291-8.

[147] Biederman, J., Fried, R., Tarko, L., Surman, C., Spencer. T., Pope, A., Grossman, R., McDermott, K., Woodworth, K. Y. & Faraone, S. V. (2014). Memantine in the

treatment of executive function deficits in adults with ADHD: a pilot-randomized double-blind controlled clinical trial. *J. Atten. Disord.*, pii: 1087054714538656.

[148] Grant, J. E., Odlaug, B. L., Schreiber, L. R., Chamberlain, S. R. & Won Kim, S. (2013). Memantine reduces stealing behavior and impulsivity in kleptomania: a pilot study. *Int. Clin. Psychopharmacol.*, *28*, 106-11.

[149] Grant, J. E., Chamberlain, S. R., Odlaug, B. L., Potenza, M. N. & Kim, S. W. (2010). Memantine shows promise in reducing gambling severity and cognitive inflexibility in pathological gambling: a pilot study. *Psychopharmacology.*, *212*, 603-12.

[150] Krupitsky, E. M., Neznanova, O., Masalov, D., Burakov, A. M., Didenko, T., Romanova, T., Tsoy, M., Bespalov, A., Slavina, T. Y., Grinenko, A. A., Petrakis, I. L., Pittman, B., Gueorguieva, R., Zvartau, E. E. & Krystal, J. H. (2007). Effect of memantine on cue-induced alcohol craving in recovering alcohol-dependent patients. *Am. J. Psychiat.*, *164*, 519-23.

[151] Bisaga, A., Comer, S., Ward, A., Popik, P., Kleber, H. & Fischman, M. (2001). The NMDA antagonist memantine attenuates the expression of opioid physical dependence in humans. *Psychopharmacology.*, *157*, 1-10.

[152] Wyss, M. T., Jolivet, R., Buck, A., Magistretti, P. J. & Weber, B. (2011). In vivo evidence for lactate as a neuronal energy source. *J. Neurosci.*, *31*, 7477-85.

[153] Pellerin, L. & Magistretti, P. J. (2011). Sweet sixteen for ANLS. *J. Cerebr. Blood. F. Met.*, *32*, 1152-66.

[154] Hutchinson, D. S., Catus, S. L., Merlin, J., Summers, R. J. & Gibbs, M. E. (2011). α_2-adrenoceptors activate noradrenalin-mediated glycogen turnover in chick astrocytes. *J. Neurochem.*, *117*, 915-26.

[155] Pase, C. S., Roversi, K., Trevizol, F., Kuhn, F. T., Dias, V. T., Roversi, K., Vey, L. T., Antoniazzi, C. T., Barcelos, R. C. & Burger, M. E. (2015). Chronic consumption of trans fat can facilitate the development of hyperactive behavior in rats. *Physiol. Behav.*, *139*, 344-50.

[156] Liso Navarro, A. A., Sikoglu, E. M., Heinze, C. R., Rogan, R. C., Russell, V. A., King, J. A. & Moore, C. M. (2014). Effect of diet on brain metabolites and behavior in spontaneously hypertensive rats. *Behav. Brain Res.*, *270*, 240-7.

[157] Heilskov Rytter, M. J., Andersen, L. B., Houmann, T., Bilenberg, N., Hvolby, A., Mølgaard, C., Michaelsen, K. F. & Lauritzen, L. (2015) Diet in the treatment of ADHD in children– a sytematic review of the literature. *Nord. J. Psychiatry.*, *69*, 1-18.

[158] Robinson, A. M. & Bucci, D. J. (2014). Individual and combined effects of physical exercise and methylphenidate on orienting behavior and social interaction in spontaneously hypertensive rats. *Behav. Neurosci.*, *128*, 703-12.

[159] Piepmeier, A. T., Shih, C. H., Whedon, M., Williams, L. M., Davis, M. E., Henning, D. A., Park, S., Calkins, S. D. & Etnier, J. L. (2015). The effect of acute exercise on cognitive performance in children with and without ADHD. *Journal of Sport and Health Science.*, *4*, 97-104.

[160] Pan, C. Y., Tsai, C. L., Chu, C. H., Sung, M. C., Huang, C. Y. & Ma, W. Y. (2015). Effects of physical exercise intervention on motor skills and executive functions n children with ADHD: a pilot study. *J. Atten. Disord.*, pii: 1087054715569282.

[161] Chuang, L. Y., Tsai, Y. J., Chang, Y. K., Huang, C. J. & Hung, T. M. (2015). Effects of acute aerobic exercise on response preparation in a Go/No Go Task in children with ADHD: An ERP study. *Journal of Sport and Health Science.*, *4*, 82-88.

In: Modulators of Glutamatergic Signaling
Editor: Zoran M. Pavlovic
ISBN: 978-1-63483-493-3

Chapter 3

METABOTROPIC GLUTAMATE RECEPTORS IN ADDICTIVE DISORDERS

***Jeremy Egnatios*[1] *and M. Foster Olive*[2,*]**
[1]School of Medicine, University of California at San Diego,
La Jolla, CA, US
[2]Department of Psychology, Arizona State University,
Tempe, AZ, US

ABSTRACT

Receptors for glutamate are primarily divided into ligand-gated ion channels (ionotropic glutamate receptors, iGluRs) and G-protein coupled receptors (metabotropic glutamate receptors, mGluRs). While ligands interacting with iGluRs alter addictive behaviors in both rodents and humans, they generally exhibit poor efficacy and side effect profiles. Alternatively, compounds that interact with mGluRs exert a more modulatory influence over glutamatergic transmission, and thus may represent a new class of pharmacotherapeutics for addictive disorders with reduced side effects. In the late 2000's, we published comprehensive reviews of preclinical studies demonstrating anti-addictive properties of mGluR ligands [1, 2]. The present chapter provides a brief introduction and historical perspective on mGluR receptors and addiction, a discussion of the co-morbidity of addictive and other neuropsychiatric disorders, a summary of preclinical studies on the anti-addictive properties of mGluR ligands that have been published from 2009-present, a brief overview of other glutamatergic medications utilized for the treatment of addictive disorders, and finally a perspective on directions for further development of mGluR ligands for the treatment of addictive disorders.

Keywords: glutamate, mGluR, metabotropic receptor, addiction, plasticity, rodent models

[*] Corresponding author: M. Foster Olive, Ph.D., Department of Psychology Arizona State University, Tempe, AZ, USA, Email: foster.olive@asu.edu.

INTRODUCTION AND HISTORICAL PERSPECTIVE

Glutamate is the most abundant excitatory neurochemical in the central nervous system and is utilized in approximately 70% of all central neuronal synapses. This excitatory amino acid mediates rapid excitatory transmission via interactions with glutamate-gated cation channels (ionotropic glutamate receptors, iGluRs) that include the N-methyl-D-aspartate (NMDA), α-amino-3-hydroxy-5-methylisoxazole-4-propionic acid (AMPA), and kainic acid (KA) receptor subtypes (see Figure 1). iGluRs are widely distributed throughout the central nervous system, but pharmacological targeting of these receptors can potentially cause severe side effects [1-4]. However, glutamate also activates metabotropic G-protein coupled receptors (mGluRs) that produce slower modulatory responses via second messenger cascades (see Figure 1). Compared to iGluRs, mGluRs are more diverse in their distribution throughout the CNS, and when targeted pharmacologically, are less prone to lead to severe side effects [1-4]. To date, eight mGluR receptor subtypes (mGluR1–8) have been identified, and are generally subdivided into one of three groups based on their sequence homology, pharmacological selectivity, and intracellular signaling mechanisms [1-4]. The Group I family of mGluRs consists of mGluR1 and mGluR5, which are primarily localized to postsynaptic elements and coupled to G_q-mediated signaling mechanisms. Group II mGluRs include mGluR2 and mGluR3, which are often found on presynaptic terminals, and since they are primarily coupled to $G_{i/o}$-mediated inhibitory signaling, they can act as classical release-regulating presynaptic autoreceptors. mGluR3 and mGluR5 are also expressed in glial cells such as astrocytes, consistent with the notion of a tripartite glutamatergic synapse (Figure 1). Finally, Group III mGluRs include mGluR4, mGluR6, mGluR7, and mGluR8, which are primarily coupled to $G_{i/o}$-mediated inhibitory signaling. It should be noted that mGluR6 is expressed exclusively in the retina and is thus generally not targeted therapeutically for neuropsychiatric disorders.

Following exocytosis, levels of extracellular glutamate are tightly regulated by sodium-dependent excitatory amino acid transporters (EAATs) designated EAAT1-5. EAAT1-3 are alternatively termed glutamate aspartate transporter (GLAST), glutamate transporter 1 (GLT-1), and excitatory amino acid carrier 1 (EAAC1), respectively. In addition to terminating glutamatergic transmission, EAATs also prevent excessive accumulation of extracellular glutamate, which can result in excitotoxicity. Glutamate is also transported from within glial cells to the extracellular milieu by the cystine-glutamate exchanger (x_c). Within glial cells, glutamate is converted to glutamine, which is transported to the presynaptic terminal for re-conversion to glutamate.

One of the most influential studies suggesting a critical role for and potential therapeutic targeting of individual mGluR subtypes in addictive disorders was published by Chiamulera and colleagues in 2001 [6]. This study demonstrated that mice carrying a targeted deletion of the mGluR5 receptor gene do not demonstrate hyperlocomotion in response to acute administration of various doses of cocaine (10, 20 or 40 mg/kg i.p.), and fail to acquire operant intravenous cocaine self-administration. However, mGluR5 null mice maintained the normal abilities to acquire operant food self-administration and demonstrated normal increases in extracellular dopamine in the ventral striatum as well as normal dopamine D_1 and D_2 receptor expression. Since this pivotal study, a large body of literature has accumulated demonstrating that not only does mGluR5 receptor deletion or pharmacological modulation alter addiction-related behaviors in rodents, but experimental manipulation of most other mGluR subtypes also influences many addictive behaviors. In the late 2000's, we published two comprehensive reviews summarizing preclinical studies confirming a role for mGluR subtypes and addictive behaviors [1, 2]. Therefore, in the present chapter, discussion of relevant preclinical studies will primarily focus on those published between 2009 and the present.

Co-Morbidity of Addictive and Other Neuropsychiatric Disorders

The phenomenon of drug addiction has previously been referred to by various colloquial and diagnostic terminologies including drug abuse or dependence, substance use or dependence, chemical dependency, to name a few. The latest (5th) revision of the Diagnostic and Statistical Manual of Mental Disorders (DSM-5) now refers to drug addiction as a "substance use disorder" (SUD), which is further subclassified by the primary abused substance (i.e., alcohol use disorder, tobacco use disorder, etc.). The prevalence of any form of addiction (behavioral or drug-related) has been estimated to range widely in a given 12-month period [7]. However, with regard to SUDs, incidence rates can vary by the primary abused substance, with alcohol abuse most common and illicit substances such heroin, cocaine, and methamphetamine substantially lower [8-10]. This disease prevalence is substantially magnified in many already burdened populations, such as those with co-occurring neuropsychiatric disorders [11, 12]. In general, patients with SUDs are seven times more likely to have one or more co-morbid non-substance related psychiatric diagnoses. Specifically, in patients with alcohol-related disorders, nearly 40% have an established co-existing mental disorder, and in patients abusing or addicted to illicit substances, the rate of psychiatric co-morbidity increases to over 50% [8].

Not only can the relationship between SUDs and other mental illnesses be dependent on the primary type of substance used, but also the co-occurring mental illness. For example, the rates of cigarette smoking in persons with schizophrenia is extremely high, ranging from 45 to nearly 90% of affected individuals [13], and substance abuse in patients with bipolar disorder are also often above 50% [14]. On the other hand, rates of alcoholism co-occurrence with major depression and anxiety disorders tend to be lower at rates of 4-13% [15, 16], although alcohol abuse is higher in patients with post-traumatic stress disorder (PTSD) [12]. The co-morbidity of SUDs and other mental illness is further complicated by the possibility, although not conclusively linked, of treatment medications actually increasing the prevalence for future psychiatric diagnoses. The most concerning of these possibilities is found in the treatment of childhood attention deficit/hyperactivity disorder (ADHD), where stimulant-based treatments are common and have been feared to increase the vulnerability to future drug use. While a possibility, not all studies support this notion. One study demonstrated that children with ADHD were at the greatest risk for developing future SUDs if left untreated, and future SUD risk was actually slightly lower in medicated patients, although SUD incidence in either group was still higher than the general population [17].

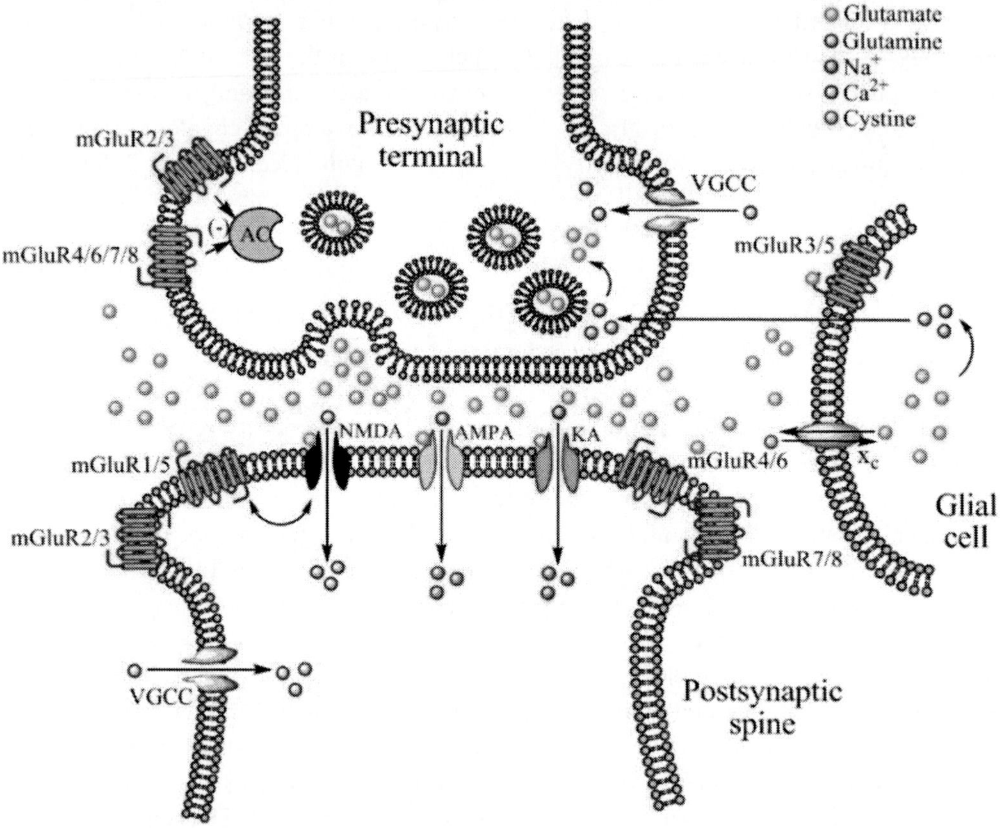

Figure 1. Diagram of the tripartite glutamatergic synapse depicting the general scheme of excitatory amino acid transmission. Reprinted with permission from [5].

There thus appears to be a strong relationship between SUDs and other neuropsychiatric disorders. However, this relationship is highly complicated by many factors including (but not limited to) specific diagnoses, primary substance used, socioeconomic disparities in SUD occurrence, and disease stage, course, and severity, in addition to any biological factors such as pharmacogenomic influences or genetic susceptibility. Further, it should be noted many experts warn against psychiatric disease evaluation in patients currently or recently recovering from substance use disorder. Often times drug effects and withdrawal symptoms can mimic other psychiatric symptomatology, and epidemiological studies may or may not sufficiently account for this possibility.

mGluRs and Drug Addiction

As noted earlier in this chapter, we published two comprehensive reviews in the late 2000's summarizing the preclinical evidence that strongly supports a role for mGluRs in mediating drug addiction and related behaviors [1, 2]. Research conducted since the publication of these reviews has not only provided additional support for this role, but recent studies have revealed details regarding the neural circuits and cellular and mechanisms involved. In addition, the recent development in selective mGluR radiotracers for use in positron emission tomography (PET) studies has enabled visualization of mGluR binding and expression levels in specific regions of the human brain in patients with SUDs. For the sake of brevity and focus, the following subsections highlight various aspects of the literature on mGluR receptors and addiction that have been published from 2009 to the present date.

Group I mGluRs

Since the seminal findings of Chiamulera et al. [6] that mice lacking mGluR5 receptors failed to demonstrate cocaine reinforcement or hyperlocomotion, the majority of studies on Group I mGluR receptors and addiction have focused on this receptor. Additional studies continue to demonstrate an ability of mGluR5 receptor antagonists to attenuate drug reward, self-administration and/or reinstatement of drug-seeking behavior for alcohol [18, 19], cocaine [20-23], methamphetamine [24-26], and opiates [22, 27, 28]. However, several studies have revealed that the ability of mGluR5 antagonists to reduce drug intake may be a result of general impairment of brain reward circuit function as revealed by intracranial self-stimulation of reductions in natural reinforcer seeking [25, 29, 30]. Despite this, there is still considerable enthusiasm for pharmacologically targeting the mGluR5 receptor for the treatment of drug addiction, and possibly co-morbid psychiatric disorders [31].

With regard to the underlying neural circuitries and mechanisms, studies employing intracranial microinjection procedures have implicated mGluR5 receptors in the central amygdala, ventral tegmental area, and dorsal and ventral striatum in regulating drug reward, intake, or seeking [32-37]. Several elegant studies have shown that eliminating mGluR5 receptors on D_1 receptor-expressing neurons in the striatum attenuates reinstatement of cocaine-seeking or escalation of alcohol intake [38, 39]. Recent evidence also suggests that striatal mGluR5 receptors form functional heteromeric complexes with other G-protein coupled receptors such as adenosine A_{2A} and dopamine D_2 receptors to regulate reward circuitry function and addictive behaviors [40].

Although most studies examining Group I mGluR receptors and their role in addiction have focused on the mGluR5 receptor, the lesser-studied mGluR1 receptor has recently gained momentum in terms of a defined role in regulating addictive behaviors [41]. In rodents and/or non-human primates, selective mGluR1 antagonists have been shown to attenuate alcohol, cocaine, and methamphetamine intake or seeking behavior [42, 43]. As with mGluR5 receptors, these effects are also observed after selective blockade of mGluR1 receptors in the nucleus accumbens and central amygdala [32, 44, 45], but other investigators have revealed a role for mGluR1 receptors in the dorsal hippocampus in cocaine-seeking [46].

In contrast to above studies, very recently it was demonstrated that positive allosteric modulation of mGluR1 receptors selectively suppresses the incubation of cocaine craving, putatively by preventing cocaine-induced accumulation of calcium-permeable AMPA receptors in the postsynaptic membrane of glutamatergic synapses onto ventral striatal medium spiny neurons [47, 48]. Given the aforementioned ability of mGluR1 antagonists to suppress the reinstatement of cocaine-seeking, the observations that mGluR1 *activation* suppresses the incubation of cocaine craving may seem counterintuitive at first glance. However, it should be emphasized that rodent models of reinstatement of drug seeking vs. cocaine craving are not necessarily isomorphic. Moreover, since both mGluR1 and mGluR5 are predominantly expressed on postsynaptic elements of glutamatergic synapses, these findings support findings of divergent modulatory effects of mGluR1 and mGluR5 on synaptic plasticity, with mGluR1 receptors primarily mediating synaptic depression, whereas mGluR5 can mediate both synaptic potentiation and depression. Thus, pharmacological alteration of synaptic plasticity in discrete brain regions may offer a therapeutic opportunity for reversing long-term neuroadaptations produced by drugs of abuse [49-54].

A highly significant advance in the field of mGluRs and addiction has been the recent development of radioligands selective for mGluR subtypes, such as [^{11}C]-ABP688 for mGluR5 receptors. This breakthrough now allows for the assessment of mGluR "status" in the healthy and diseased brain. Although only a few studies have been conducted, evidence thus far suggests that striatal mGluR5 binding is reduced by ~20-40% patients with a history of cocaine addiction or long-term cigarette smoking [55-57]. Interestingly, a very recent study showed that long-term abstinence from cigarette smoking was associated with a normalization of striatal mGluR5 binding to those of control subjects, suggesting considerable plasticity in mGluR5 expression in the addicted and recovering brain. PET imaging of mGluR5 binding has also been conducted in patients with major depressive disorder and [58], supporting the utility of monitoring glutamate homeostasis in SUD patients with or without co-morbid psychiatric diagnoses. The development of PET ligands selective for mGluR1 has also been recently reported [59, 60], opening new avenues for research into this lesser studied Group I mGluR in addictive disorders.

Group II mGluRs

Group II mGluRs (mGluR2 and mGluR3) are predominantly found presynaptically on glutamatergic terminals and serve as release-regulating autoreceptors in numerous regions including brain reward circuitry. A major hypothesis put forth by Kalivas and colleagues [61, 62] on the role of glutamate in addiction asserts that chronic drug use produces profound adaptations in prefrontal-striatal glutamatergic projections. These adaptations result in reductions in extracellular glutamate and thus loss of glutamatergic tone on presynaptic Group II mGluR receptors, which ultimately leads to loss of top-down control of drug seeking by executive prefrontal mechanisms. Thus, a reasonable and testable hypothesis is that enhancing or promoting Group II mGluR activity in the nucleus accumbens, effectively blunting downstream corticostriatal glutamate and mesolimbic dopamine and may reduce drug-seeking behavior. This has been shown to be the case, with agonists at group II mGluRs suppressing drug intake, reward, and reinstatement for cocaine [63-66], methamphetamine [67, 68], alcohol [69-71] and nicotine [72]. Microinjection studies have implicated mGluR2/3 receptors in the ventral tegmental area, nucleus accumbens, and prefrontal cortex in mediating these effects [63, 73]. Recent studies with newly developed ligands selective for mGluR2 over mGluR3 receptors have shown similar effects [74-76]. However, in light of this, a potential role for mGluR3 receptors in addictive processes should not be overlooked, as the Grm3 gene, which encodes the mGluR3 receptor, has been shown to be a candidate gene for susceptibility to alcoholism [77]. Finally, while PET radioligands for selective for Group II mGluRs have only recently begun to be developed [78], studies in non-human primates have shown that striatal mGluR2/3 receptors are significantly up-regulated in cocaine addiction [79]. Thus, the functional status of mGluR2/3 receptors in SUD patients remains to be determined.

Group III mGluRs

Similar to Group II mGluR receptors, centrally expressed Group III mGluRs (mGluR4, mGluR7 and mGluR8) are primarily expressed on presynaptic elements of glutamatergic synapses. The most frequently studied Group III mGluR in the context of drug addiction is the mGluR7 receptor. As with mGluR2/3 receptors, activation of mGluR7 has been shown to reduce alcohol consumption [80] and inhibit the reinstatement of cocaine-seeking behavior [81]. Other studies in rodents utilizing gene mapping and mRNA segregating methods in congenic mice have identified mGluR7 as a cis-regulated gene for alcohol consumption [82]. Genome wide associated studies (GWAS) in humans have also implicated mGluR7 in addiction, with multiple single nucleotide polymorphisms found in significantly higher frequency in alcoholics as compared to the general population [83].

Only a handful of studies have implicated either mGluR4 or mGluR8 receptors in addiction, primarily due to the lack of availability of selective ligands for individual Group III mGluR subtypes [84]. Over a decade ago, it was demonstrated that mice carrying a targeted deletion of the gene encoding mGluR4 do not show locomotor stimulant properties of alcohol [85]. However, the recent development of a selective mGluR4 agonists and positive allosteric modulators have revealed that activation of mGluR4 decreases the expression of cocaine- but not nicotine-induced locomotor sensitization [86].

With regards to mGluR8, it was demonstrated that a putative mGluR8 agonist suppressed alcohol intake and cue-induced reinstatement of alcohol-seeking [69]. Acute exposure to psychostimulants such as cocaine decreases mGluR8 expression down regulate protein expression of this receptor, while chronic exposure produces the opposite effects [84]. The gene encoding mGluR8 has also been associated with vulnerability to heroin addiction in humans [87]. Given the paucity of published studies on Group III mGluRs and addiction, clearly there is a need for additional investigations into the therapeutic potential of these receptors for addiction treatment.

Current Glutamatergic Medications for the Treatment of Addictive Disorders

With an established role for glutamate in addictive disorders, it is disappointing that only a handful of medications with known glutamatergic activity have been proposed and tested for the treatment of addiction. Perhaps even more disappointing is that none of the glutamatergic medications currently approved for use by the U.S. Food and Drug Administration (FDA) have been specifically approved for the treatment of addictive disorders.

Nonetheless, below is a summary of various medications with glutamatergic mechanisms of action that are approved by the FDA for other medical conditions that have shown some degree of promise for the treatment of addictive disorders. For more detailed descriptions, the reader is referred to recent reviews published elsewhere [88-90].

N-Acetylcysteine

N-Acetylcysteine (NAC) is a cystine pro-drug most commonly used as a mucolytic agent and to treat acetaminophen toxicity. Following ingestion, NAC is bioconverted to cysteine, which binds to the cystine-glutamate exchanger (x_c; see Figure 1) to promote glutathione synthesis. In addition, extracellular cysteine is taken up by glial cells in exchange for the release of glutamate, thus increasing extracellular glutamate levels. NAC has been demonstrated in multiple animal studies to normalize glutamate homeostasis during withdrawal, when extracellular glutamate levels are usually reduced, to produce reductions in withdrawal symptoms and/or reinstatement. This has been demonstrated to be the case for multiple drugs of abuse, including nicotine, marijuana, heroin, nicotine, and cocaine (reviewed elsewhere [91-93]). NAC has been shown to reduce either drug craving, cue reactivity, or abstinence rates, as well as normalize glutamate homeostasis in brain regions such as the anterior cingulate cortex in dependent individuals.

D-cycloserine

D-cycloserine (DCS) is an enantiomer specific version of the anti-tuberculosis antibiotic seromycin. DCS was found to be a co-agonist at the NMDA receptor, and is thus able to enhance NMDA receptor activity and increase calcium influx into neurons. As a result of this action, DCS is believed to enhance synaptic plasticity and thus facilitate the extinction of drug-seeking or the salience of drug-cue associations as normally attempted through cue exposure therapy. Multiple studies have shown DCS as efficacious at promoting extinction and reducing reinstatement of cocaine seeking as well as fear extinction in animal models [94]. However the clinical efficacy of DCS as a treatment for addiction has been mixed, with some studies showing DCS-induced reduced drug cue reactivity in alcohol dependent patients [95, 96], others showing no effect in alcoholics or cigarette smokers [97, 98], and still others showing unexpected increase in drug cue reactivity or craving in cocaine-dependent individuals [99-101]. Thus, the potential therapeutic benefits of DCS in facilitating extinction and reducing cue-evoked craving may be limited to the treatment addiction to alcohol but not other abused drugs.

Modafinil

Modafinil is a wake-promoting agent approved for the treatment of narcolepsy and sleep-wake disturbance in shift work. Multiple neurochemical mechanisms of modafinil action have been proposed, including potentiation of monoaminergic transmission and reducing GABA transmission, as well as modulating extracellular glutamate levels [102]. In rodents, modafinil attenuates the reinstatement of cocaine- and opiate-seeking behavior via glutamatergic mechanisms [103, 104]. In humans, modafinil has been shown to produce beneficial effects in reducing cocaine craving and promoting abstinence [105-107], but its efficacy for other SUDs has yet to be established. Thus, modafinil may have some but limited potential for treating SUDs.

Memantine

Memantine is approved for the treatment of Alzheimer's disease, where it has been thought to reduce excitotoxicity via its actions as an NMDA antagonist. Additionally, it has been found to have some anticholinergic and antiserotonergic properties. It has been demonstrated in rodents that memantine attenuates self-administration of alcohol and nicotine but not cocaine [108, 109], and attenuates sensitization to the rewarding effects of morphine [110]. Clinical studies have reported that memantine reduced intake of and craving for alcohol [111], yet another report indicated that memantine worsens abstinence rates in actively drinking alcoholics [112]. Memantine has been reported to attenuate the subjective effects of heroin and cigarette smoking [113, 114]. Thus, there is some but relatively limited evidence for the potential for memantine for SUD treatment.

Anticonvulsants

There is an increasing body of literature indicating that anticonvulsant medications such as topiramate, gabapentin, pregabalin, or lamotrigine may be of some use in the treatment of addiction. These anticonvulsant medications have multiple modes of action, including inhibition of presynaptic of voltage-gated ion channels to attenuate the release of various neurotransmitters including glutamate. Some anticonvulsants such as topiramate have direct effects on iGluRs such as GluR5-containing kainate receptors. There are numerous studies from both the animal and human literature suggesting that anticonvulsants may attenuate the intake of various drugs of abuse such as alcohol and methamphetamine, and particularly effective for the relief of alcohol withdrawal symptoms [2, 88, 115, 116]. However, with regard to cocaine, initial clinical trials provided hope that anticonvulsants may represent a new class of medications for the treatment of cocaine addiction. Yet the Cochrane Library recently published a pooled meta-analysis and review of numerous studies examining the clinical efficacy of anticonvulsants for cocaine addiction, and provided the general conclusion that there are no significant evidence that anticonvulsants reduce cocaine intake, craving, or relapse rates [117].

Future Directions for the Development of mGluR Ligands for the Treatment of Addictive Disorders

At the beginning of the 21st century, the prospect of advancing mGluR ligands into testing in human subjects to assess their efficacy in treating SUDs seemed like a distant hope. However, this goal is now potentially achievable in the very near future, since various mGluR ligands have entered clinical trials at various phases. Unfortunately, none of these trials are designed with addictive disorders as the primary medical indication. Nonetheless, it is worth briefly reviewing the results of clinical trials utilizing mGluR-acting compounds that have been published thus far.

The first clinical trials involving mGluR ligands were actually conducted more than 30 years ago, without the knowledge that the compound being tested was in fact an mGluR ligand, and even more ironically prior to the actual cloning and characterization of any mGluR subtype. These clinical studies emerged in the late 1970s and early 1980s and revealed that the compound 1-(3-chlorophenyl)-3-(3-methyl-5-oxo-4H-imidazol-2-yl)urea, more frequently referred to as fenobam, exhibited anxiolytic effects in humans, despite it have a non-benzodiazepine structure [118-122]. However, clinical trials were soon discontinued following reports of adverse dose-related side effects including dizziness, paresthesias, sedation, and depersonalization. Some two decades later, it was revealed that fenobam is actually a selective negative allosteric modulator of mGluR5 receptors [123, 124]. Given the established role of these receptors in regulating aberrant synaptic plasticity and protein synthesis in Fragile X syndrome, initial trials in this patient population with lower doses of fenobam produced cognitive improvement with few CNS-related adverse side effects [125]. Additional clinical trials of various other mGluR5 negative allosteric modulators, including AFQ056, RG7090, and STX-107 have also been conducted, but unfortunately many of these trials have failed due to poor clinical efficacy [126]. Still, mGluR5 inhibitors such as ADX10059, basimglurant/RO4917523, and mavoglurant continue to be developed for other medical conditions including gastroesophageal reflux disease, anxiety, depression and migraine, and at proper doses appear to be generally well tolerated [127-131]. Ligands acting on other mGluR subtypes have also advanced to clinical trials, including the mGluR2/3 agonist LY354740 for anxiety [132-135] and the mGluR2/3 agonist pro-drug LY2140023/ pomeglumetad in the treatment of schizophrenia [136-139]. Thus, there are several mGluR2/3 and mGluR5 ligands that could potentially be tested in patients with SUDs, and such trials are eagerly awaited given the substantial body of preclinical evidence implicating these receptors in addiction to many abused substances.

References

[1] Olive, M. F. (2009). Metabotropic glutamate receptor ligands as potential therapeutics for drug addiction. *Curr. Drug Abuse Rev.*, *2*, 83-98.

[2] Gass, J. T. & Olive, M. F. (2008). Glutamatergic substrates of drug addiction and alcoholism. Biochem. *Pharmacol.*, *75*, 218-265.

[3] Niswender, C. M. & Conn, P. J. (2010). Metabotropic glutamate receptors: physiology, pharmacology, and disease. *Annu. Rev. Pharmacol. Toxicol.*, *50*, 295-322.

[4] Conn, P. J., Lindsley, C. W., Meiler, J. & Niswender, C. M. (2014). Opportunities and challenges in the discovery of allosteric modulators of GPCRs for treating CNS disorders. *Nat. Rev. Drug Discov.*, *13*, 692-708.

[5] Nemirovsky, N. E. & Olive, M. F. Medications for the treatment of cocaine addiction: focus on glutamatergic compounds In: Fang X.C., Yue L., editors. *Cocaine Abuse: Pharmacology, Treatment and Relapse Prevention*. Hauppauge, NY: Nova Biomedical, 2012.115-129.

[6] Chiamulera, C., Epping-Jordan, M. P., Zocchi, A., Marcon, C., Cottiny, C., Tacconi, S., Corsi, M., Orzi, F. & Conquet, F. (2001). Reinforcing and locomotor stimulant effects of cocaine are absent in mGluR5 null mutant mice. *Nat. Neurosci.*, *4*, 873-4.

[7] Sussman, S., Lisha, N. & Griffiths, M. (2011). Prevalence of the addictions: a problem of the majority or the minority? *Eval. Health Prof.*, *34*, 3-56.

[8] Regier, D. A., Farmer, M. E., Rae, D. S., Locke, B. Z., Keith, S. J., Judd, L. L. & Goodwin, F. K. (1990). Comorbidity of mental disorders with alcohol and other drug abuse. Results from the Epidemiologic Catchment Area (ECA) Study. *JAMA*, *264*, 2511-8.

[9] Kessler, R. C., Berglund, P., Demler, O., Jin, R., Merikangas, K. R. & Walters, E. E. (2005). Lifetime prevalence and age-of-onset distributions of DSM-IV disorders in the National Comorbidity Survey Replication. *Arch. Gen. Psychiatry*, *62*, 593-602.

[10] Schulden, J. D., Thomas, Y. F. & Compton, W. M. (2009). Substance abuse in the United States: findings from recent epidemiologic studies. *Curr. Psychiatry Rep.*, *11*, 353-9.

[11] MacDonald, B. K., Cockerell, O. C., Sander, J. W. & Shorvon, S. D. (2000). The incidence and lifetime prevalence of neurological disorders in a prospective community-based study in the UK. *Brain*, *123 (Pt 4)*, 665-76.

[12] Brady, K. T. & Sinha, R. (2005). Co-occurring mental and substance use disorders: the neurobiological effects of chronic stress. *Am. J. Psychiatry*, *162*, 1483-93.

[13] Wing, V. C., Wass, C. E., Soh, D. W. & George, T. P. (2012). A review of neurobiological vulnerability factors and treatment implications for comorbid tobacco dependence in schizophrenia. *Ann. N. Y. Acad. Sci.*, *1248*, 89-106.

[14] Quello, S. B., Brady, K. T. & Sonne, S. C. (2005). Mood disorders and substance use disorder: a complex comorbidity. *Sci. Pract. Perspect.*, *3*, 13-21.

[15] Hasin, D. S., Goodwin, R. D., Stinson, F. S. & Grant, B. F. (2005). Epidemiology of major depressive disorder: results from the National Epidemiologic Survey on Alcoholism and Related Conditions. *Arch. Gen. Psychiatry*, *62*, 1097-106.

[16] Grant, B. F., Hasin, D. S., Stinson, F. S., Dawson, D. A., June Ruan, W., Goldstein, R. B., Smith, S. M., Saha, T. D. & Huang, B. (2005). Prevalence, correlates, co-morbidity, and comparative disability of DSM-IV generalized anxiety disorder in the USA: results from the National Epidemiologic Survey on Alcohol and Related Conditions. *Psychol. Med.*, *35*, 1747-59.

[17] Wilens, T. E., Faraone, S. V., Biederman, J. & Gunawardene, S. (2003). Does stimulant therapy of attention-deficit/hyperactivity disorder beget later substance abuse? A meta-analytic review of the literature. *Pediatrics*, *111*, 179-85.

[18] Cozzoli, D. K., Strong-Kaufman, M. N., Tanchuck, M. A., Hashimoto, J. G., Wiren, K. M. & Finn, D. A. (2014). The effect of mGluR5 antagonism during binge drinking on subsequent ethanol intake in C57BL/6J mice: sex- and age-induced differences. *Alcohol. Clin. Exp. Res.*, *38*, 730-738.

[19] Kotlinska, J. H., Bochenski, M. & Danysz, W. (2011). The role of group I mGlu receptors in the expression of ethanol-induced conditioned place preference and ethanol withdrawal seizures in rats. *Eur. J. Pharmacol.*, *670*, 154-61.

[20] Keck, T. M., Yang, H. J., Bi, G. H., Huang, Y., Zhang, H. Y., Srivastava, R., Gardner, E. L., Newman, A. H. & Xi, Z. X. (2013). Fenobam sulfate inhibits cocaine-taking and cocaine-seeking behavior in rats: implications for addiction treatment in humans. *Psychopharmacology (Berl).*, *229*, 253-65.

[21] Keck, T. M., Zou, M. F., Bi, G. H., Zhang, H. Y., Wang, X. F., Yang, H. J., Srivastava, R., Gardner, E. L., Xi, Z. X. & Newman, A. H. (2014). A novel mGluR5 antagonist, MFZ 10-7, inhibits cocaine-taking and cocaine-seeking behavior in rats. *Addict. Biol.*, *19*, 195-209.

[22] Veeneman, M. M., Boleij, H., Broekhoven, M. H., Snoeren, E. M., Guitart, M. M., Cousijn, J., Spooren, W. & Vanderschuren, L. J. (2011). Dissociable roles of mGlu5 and dopamine receptors in the rewarding and sensitizing properties of morphine and cocaine. *Psychopharmacology (Berl).*, *214*, 863-76.

[23] Wang, X., Moussawi, K., Knackstedt, L., Shen, H. & Kalivas, P. W. (2013). Role of mGluR5 neurotransmission in reinstated cocaine-seeking. *Addict. Biol.*, *18*, 40-9.

[24] Gass, J. T., Osborne, M. P., Watson, N. L., Brown, J. L. & Olive, M. F. (2009). mGluR5 antagonism attenuates methamphetamine reinforcement and prevents reinstatement of methamphetamine-seeking behavior in rats. *Neuropsychopharmacology*, *34*, 820-33.

[25] Watterson, L. R., Kufahl, P. R., Nemirovsky, N. E., Sewalia, K., Hood, L. E. & Olive, M. F. (2013). Attenuation of reinstatement of methamphetamine-, sucrose-, and food-seeking behavior in rats by fenobam, a metabotropic glutamate receptor 5 negative allosteric modulator. *Psychopharmacology (Berl).*, *225*, 151-9.

[26] Chesworth, R., Brown, R. M., Kim, J. H. & Lawrence, A. J. (2013). The metabotropic glutamate 5 receptor modulates extinction and reinstatement of methamphetamine-seeking in mice. *PLoS One*, *8*, e68371.

[27] Brown, R. M., Stagnitti, M. R., Duncan, J. R. & Lawrence, A. J. (2012). The mGlu5 receptor antagonist MTEP attenuates opiate self-administration and cue-induced opiate-seeking behaviour in mice. *Drug Alcohol. Depend.*, *123*, 264-268.

[28] Manzanedo, C., Aguilar, M. A., Minarro, J. & Rodriguez-Arias, M. (2011). Effects of CNQX and MPEP on sensitization to the rewarding effects of morphine. *Eur. J. Pharmacol.*, *654*, 42-6.

[29] Kenny, P. J., Boutrel, B., Gasparini, F., Koob, G. F. & Markou, A. (2005). Metabotropic glutamate 5 receptor blockade may attenuate cocaine self-administration by decreasing brain reward function in rats. *Psychopharmacology (Berl).*, *179*, 247-54.

[30] Cleva, R. M., Watterson, L. R., Johnson, M. A. & Olive, M. F. (2012). Differential modulation of thresholds for intracranial self-stimulation by mGlu5 positive and negative allosteric modulators: implications for effects on drug self-administration. *Front. Pharmacol.*, *2*, 93 (1-7).

[31] Terbeck, S., Akkus, F., Chesterman, L. P. & Hasler, G. (2015). The role of metabotropic glutamate receptor 5 in the pathogenesis of mood disorders and addiction: combining preclinical evidence with human Positron Emission Tomography (PET) studies. *Front. Neurosci.*, *9*, 86.

[32] Cozzoli, D. K., Courson, J., Wroten, M. G., Greentree, D. I., Lum, E. N., Campbell, R. R., Thompson, A. B., Maliniak, D., Worley, P. F., Jonquieres, G., Klugmann, M., Finn, D. A. & Szumlinski, K. K. (2014). Binge alcohol drinking by mice requires intact Group 1 metabotropic glutamate receptor signaling within the central nucleus of the amygdale. *Neuropsychopharmacology*, *39*, 435-44.

[33] Cozzoli, D. K., Courson, J., Caruana, A. L., Miller, B. W., Greentree, D. I., Thomspon, A. B., Wroten, M. G., Zhang, P. W., Xiao, B., Hu, J. H., Klugmann, M., Metten, P., Worley, P. F., Crabbe, J. C. & Szumlinski, K. K. (2012). Nucleus accumbens mGluR5-associated signaling regulates binge alcohol drinking under drinking-in-the-dark procedures. *Alcohol. Clin. Exp. Res.*, *36*, 1623-1633.

[34] D'Souza, M. S. & Markou, A. (2011). Metabotropic glutamate receptor 5 antagonist 2-methyl-6-(phenylethynyl)pyridine (MPEP) microinfusions into the nucleus accumbens

shell or ventral tegmental area attenuate the reinforcing effects of nicotine in rats. *Neuropharmacology*, *61*, 1399-405.

[35] Knackstedt, L. A., Trantham-Davidson, H. L. & Schwendt, M. (2014). The role of ventral and dorsal striatum mGluR5 in relapse to cocaine-seeking and extinction learning. *Addict. Biol.*, *19*, 87-101.

[36] Sinclair, C. M., Cleva, R. M., Hood, L. E., Olive, M. F. & Gass, J. T. (2012). mGluR5 receptors in the basolateral amygdala and nucleus accumbens regulate cue-induced reinstatement of ethanol-seeking behavior. *Pharmacol. Biochem. Behav.*, *101*, 329-35.

[37] Roohi, N., Sarihi, A., Shahidi, S., Zarei, M. & Haghparast, A. (2014). Microinjection of the mGluR5 antagonist MTEP into the nucleus accumbens attenuates the acquisition but not expression of morphine-induced conditioned place preference in rats. Pharmacol. *Biochem. Behav.*, *126C*, 109-115.

[38] Novak, M., Halbout, B., O'Connor, E. C., Rodriguez Parkitna, J., Su, T., Chai, M., Crombag, H. S., Bilbao, A., Spanagel, R., Stephens, D. N., Schutz, G. & Engblom, D. (2010). Incentive learning underlying cocaine-seeking requires mGluR5 receptors located on dopamine D1 receptor-expressing neurons. *J. Neurosci.*, *30*, 11973-82.

[39] Parkitna, J. R., Sikora, M., Golda, S., Golembiowska, K., Bystrowska, B., Engblom, D., Bilbao, A. & Przewlocki, R. (2013). Novelty-seeking behaviors and the escalation of alcohol drinking after abstinence in mice are controlled by metabotropic glutamate receptor 5 on neurons expressing dopamine D_1 receptors. *Biol. Psychiatry*, *73*, 263-70.

[40] Brown, R. M., Mustafa, S., Ayoub, M. A., Dodd, P. R., Pfleger, K. D. & Lawrence, A. J. (2012). mGlu5 receptor functional interactions and addiction. *Front. Pharmacol.*, *3*, 84.

[41] Lesage, A. & Steckler, T. (2010). Metabotropic glutamate mGlu1 receptor stimulation and blockade: therapeutic opportunities in psychiatric illness. *Eur. J. Pharmacol.*, *639*, 2-16.

[42] Achat-Mendes, C., Platt, D. M. & Spealman, R. D. (2012). Antagonism of metabotropic glutamate 1 receptors attenuates behavioral effects of cocaine and methamphetamine in squirrel monkeys. *J. Pharmacol. Exp. Ther.*, *343*, 214-24.

[43] Lominac, K. D., Kapasova, Z., Hannun, R. A., Patterson, C., Middaugh, L. D. & Szumlinski, K. K. (2006). Behavioral and neurochemical interactions between group I mGluR antagonists and ethanol: potential insight into their anti-addictive properties. *Drug Alcohol Depend.*, *85*, 142-156.

[44] Lum, E. N., Campbell, R. R., Rostock, C. & Szumlinski, K. K. (2014). mGluR1 within the nucleus accumbens regulates alcohol intake in mice under limited-access conditions. *Neuropharmacology*, *79*, 679-87.

[45] Xie, X., Lasseter, H. C., Ramirez, D. R., Ponds, K. L., Wells, A. M. & Fuchs, R. A. (2012). Subregion-specific role of glutamate receptors in the nucleus accumbens on drug context-induced reinstatement of cocaine-seeking behavior in rats. *Addict. Biol.*, *17*, 287-99.

[46] Xie, X., Ramirez, D. R., Lasseter, H. C. & Fuchs, R. A. (2010). Effects of mGluR1 antagonism in the dorsal hippocampus on drug context-induced reinstatement of cocaine-seeking behavior in rats. *Psychopharmacology (Berl).*, *208*, 1-11.

[47] Loweth, J. A., Scheyer, A. F., Milovanovic, M., LaCrosse, A. L., Flores-Barrera, E., Werner, C. T., Li, X., Ford, K. A., Le, T., Olive, M. F., Szumlinski, K. K., Tseng, K. Y.

& Wolf, M. E. (2014). Synaptic depression via mGluR1 positive allosteric modulation suppresses cue-induced cocaine craving. *Nat. Neurosci.*, *17*, 73-80.

[48] Loweth, J. A., Tseng, K. Y. & Wolf, M. E. (2013). Using metabotropic glutamate receptors to modulate cocaine's synaptic and behavioral effects: mGluR1 finds a niche. *Curr. Opin. Neurobiol.*, *23*, 500-6.

[49] Bird, M. K. & Lawrence, A. J. (2009). Group I metabotropic glutamate receptors: involvement in drug-seeking and drug-induced plasticity. *Curr. Mol. Pharmacol.*, *2*, 83-94.

[50] Bellone, C. & Mameli, M. (2012). mGluR-dependent synaptic plasticity in drug-seeking. *Front. Pharmacol.*, *3*, 159:1-6.

[51] Huang, C. C., Yeh, C. M., Wu, M. Y., Chang, A. Y., Chan, J. Y., Chan, S. H. & Hsu, K. S. (2011). Cocaine withdrawal impairs metabotropic glutamate receptor-dependent long-term depression in the nucleus accumbens. *J. Neurosci.*, *31*, 4194-203.

[52] Luscher, C. (2013). Drug-evoked synaptic plasticity causing addictive behavior. *J. Neurosci.*, *33*, 17641-6.

[53] Luscher, C. & Huber, K. M. (2010). Group 1 mGluR-dependent synaptic long-term depression: mechanisms and implications for circuitry and disease. *Neuron*, *65*, 445-59.

[54] Luscher, C. & Malenka, R. C. (2011). Drug-evoked synaptic plasticity in addiction: from molecular changes to circuit remodeling. *Neuron*, *69*, 650-63.

[55] Martinez, D., Slifstein, M., Nabulsi, N., Grassetti, A., Urban, N. B., Perez, A., Liu, F., Lin, S. F., Ropchan, J., Mao, X., Kegeles, L. S., Shungu, D. C., Carson, R. E. & Huang, Y. (2014). Imaging glutamate homeostasis in cocaine addiction with the metabotropic glutamate receptor 5 positron emission tomography radiotracer [^{11}C]ABP688 and magnetic resonance spectroscopy. *Biol. Psychiatry*, *75*, 165-71.

[56] Akkus, F., Ametamey, S. M., Treyer, V., Burger, C., Johayem, A., Umbricht, D., Gomez Mancilla, B., Sovago, J., Buck, A. & Hasler, G. (2013). Marked global reduction in mGluR5 receptor binding in smokers and ex-smokers determined by [^{11}C]ABP688 positron emission tomography. *Proc. Natl. Acad. Sci. U. S. A.*, *110*, 737-42.

[57] Milella, M. S., Marengo, L., Larcher, K., Fotros, A., Dagher, A., Rosa-Neto, P., Benkelfat, C. & Leyton, M. (2014). Limbic system mGluR5 availability in cocaine dependent subjects: a high-resolution PET [^{11}C]ABP688 study. *Neuroimage*, *98*, 195-202.

[58] Deschwanden, A., Karolewicz, B., Feyissa, A. M., Treyer, V., Ametamey, S. M., Johayem, A., Burger, C., Auberson, Y. P., Sovago, J., Stockmeier, C. A., Buck, A. & Hasler, G. (2011). Reduced metabotropic glutamate receptor 5 density in major depression determined by ^{11}C-ABP688 PET and postmortem study. *Am. J. Psychiatry*, *168*, 727-34.

[59] Toyohara, J., Sakata, M., Fujinaga, M., Yamasaki, T., Oda, K., Ishii, K., Zhang, M. R., Moriguchi, J. C. M. & Ishiwata, K. (2013). Preclinical and the first clinical studies on ^{11}C-ITMM for mapping metabotropic glutamate receptor subtype 1 by positron emission tomography. *Nucl. Med. Biol.*, *40*, 214-20.

[60] Toyohara, J., Sakata, M., Oda, K., Ishii, K., Ito, K., Hiura, M., Fujinaga, M., Yamasaki, T., Zhang, M. R. & Ishiwata, K. (2013). Initial human PET studies of metabotropic glutamate receptor type 1 ligand ^{11}C-ITMM. *J. Nucl. Med.*, *54*, 1302-7.

[61] Kalivas, P. W. (2009). The glutamate homeostasis hypothesis of addiction. *Nat. Rev. Neurosci.*, *10*, 561-72.

[62] Moussawi, K. & Kalivas, P. W. (2010). Group II metabotropic glutamate receptors (mGlu2/3) in drug addiction. *Eur. J. Pharmacol.*, *639*, 115-22.

[63] Peters, J. & Kalivas, P. W. (2006). The group II metabotropic glutamate receptor agonist, LY379268, inhibits both cocaine- and food-seeking behavior in rats. *Psychopharmacology (Berl).*, *186*, 143-9.

[64] Baptista, M. A. S., Martin-Fardon, R. & Weiss, F. (2004). Preferential effects of the metabotropic glutamate 2/3 receptor agonist LY379268 on conditioned reinstatement versus primary reinforcement: Comparison between cocaine and a potent conventional reinforcer. *J. Neurosci.*, *24*, 4723-4727.

[65] Cannella, N., Halbout, B., Uhrig, S., Evrard, L., Corsi, M., Corti, C., Deroche-Gamonet, V., Hansson, A. C. & Spanagel, R. (2013). The mGluR2/3 agonist LY379268 induced anti-reinstatement effects in rats exhibiting addiction-like behavior. *Neuropsychopharmacology*, *38*, 2048-56.

[66] Dhanya, R. P., Sheffler, D. J., Dahl, R., Davis, M., Lee, P. S., Yang, L., Nickols, H. H., Cho, H. P., Smith, L. H., D'Souza, M. S., Conn, P. J., Der-Avakian, A., Markou, A. & Cosford, N. D. (2014). Design and synthesis of systemically active metabotropic glutamate subtype-2 and -3 (mGlu2/3) receptor positive allosteric modulators (PAMs): pharmacological characterization and assessment in a rat model of cocaine dependence. *J. Med. Chem.*, *57*, 4154-72.

[67] Crawford, J. T., Roberts, D. C. & Beveridge, T. J. (2013). The group II metabotropic glutamate receptor agonist, LY379268, decreases methamphetamine self-administration in rats. *Drug Alcohol Depend.*, *132*, 414-419.

[68] Kufahl, P. R., Watterson, L. R., Nemirovsky, N. E., Hood, L. E., Villa, A., Halstengard, C., Zautra, N. & Olive, M. F. (2013). Attenuation of methamphetamine seeking by the mGluR2/3 agonist LY379268 in rats with histories of restricted and escalated self-administration. *Neuropharmacology*, *66*, 290-301.

[69] Backstrom, P. & Hyytia, P. (2005). Suppression of alcohol self-administration and cue-induced reinstatement of alcohol seeking by the mGlu2/3 receptor agonist LY379268 and the mGlu8 receptor agonist (S)-3,4-DCPG. *Eur. J. Pharmacol.*, *528*, 110-118.

[70] Kufahl, P. R., Martin-Fardon, R. & Weiss, F. (2011). Enhanced sensitivity to attenuation of conditioned reinstatement by the mGluR 2/3 agonist LY379268 and increased functional activity of mGluR 2/3 in rats with a history of ethanol dependence. *Neuropsychopharmacology*, *36*, 2762-73.

[71] Sidhpura, N., Weiss, F. & Martin-Fardon, R. (2010). Effects of the mGlu2/3 agonist LY379268 and the mGlu5 antagonist MTEP on ethanol seeking and reinforcement are differentially altered in rats with a history of ethanol dependence. *Biol. Psychiatry*, *67*, 804-11.

[72] Liechti, M. E., Lhuillier, L., Kaupmann, K. & Markou, A. (2007). Metabotropic glutamate 2/3 receptors in the ventral tegmental area and the nucleus accumbens shell are involved in behaviors relating to nicotine dependence. *J. Neurosci.*, *27*, 9077-9085.

[73] Lu, L., Xue, Y., Steketee, J. D., Rebec, G. V. & Sun, W. (2012). Regulation of cocaine-induced reinstatement by group II metabotropic glutamate receptors in the ventral tegmental area. *Psychopharmacology (Berl).*, *220*, 75-85.

[74] Dhanya, R. P., Sidique, S., Sheffler, D. J., Nickols, H. H., Herath, A., Yang, L., Dahl, R., Ardecky, R., Semenova, S., Markou, A., Conn, P. J. & Cosford, N. D. (2011). Design and synthesis of an orally active metabotropic glutamate receptor subtype-2 (mGluR2) positive allosteric modulator (PAM) that decreases cocaine self-administration in rats. *J. Med. Chem.*, *54*, 342-353.

[75] in, X., Semenova, S., Yang, L., Ardecky, R., Sheffler, D. J., Dahl, R., Conn, P. J., Cosford, N. D. & Markou, A. (2010). The mGluR2 positive allosteric modulator BINA decreases cocaine self-administration and cue-induced cocaine-seeking and counteracts cocaine-induced enhancement of brain reward function in rats. *Neuropsychopharmacology*, *35*, 2021-36.

[76] Meinhardt, M. W., Hansson, A. C., Perreau-Lenz, S., Bauder-Wenz, C., Stahlin, O., Heilig, M., Harper, C., Drescher, K. U., Spanagel, R. & Sommer, W. H. (2013). Rescue of infralimbic mGluR2 deficit restores control over drug-seeking behavior in alcohol dependence. *J. Neurosci.*, *33*, 2794-806.

[77] Levey, D. F., Le-Niculescu, H., Frank, J., Ayalew, M., Jain, N., Kirlin, B., Learman, R., Winiger, E., Rodd, Z., Shekhar, A., Schork, N., Kiefe, F., Wodarz, N., Muller-Myhsok, B., Dahmen, N., Nothen, M., Sherva, R., Farrer, L., Smith, A. H., Kranzler, H. R., Rietschel, M., Gelernter, J. & Niculescu, A. B. (2014). Genetic risk prediction and neurobiological understanding of alcoholism. *Transl. Psychiatry*, *4*, e391.

[78] Sobrio, F. (2013). Radiosynthesis of carbon-11 and fluorine-18 labelled radiotracers to image the ionotropic and metabotropic glutamate receptors. *J. Labelled Comp. Radiopharm.*, *56*, 180-6.

[79] Beveridge, T. J., Smith, H. R., Nader, M. A. & Porrino, L. J. (2011). Group II metabotropic glutamate receptors in the striatum of non-human primates: dysregulation following chronic cocaine self-administration. *Neurosci. Lett.*, *496*, 15-9.

[80] Bahi, A., Fizia, K., Dietz, M., Gasparini, F. & Flor, P. J. (2012). Pharmacological modulation of mGluR7 with AMN082 and MMPIP exerts specific influences on alcohol consumption and preference in rats. *Addict. Biol.*, *17*, 235-47.

[81] Li, X., Li, J., Gardner, E. L. & Xi, Z. X. (2010). Activation of mGluR7s inhibits cocaine-induced reinstatement of drug-seeking behavior by a nucleus accumbens glutamate-mGluR2/3 mechanism in rats. *J. Neurochem.*, *114*, 1368-80.

[82] Vadasz, C., Saito, M., Gyetvai, B. M., Oros, M., Szakall, I., Kovacs, K. M., Prasad, V. V. & Toth, R. (2007). Glutamate receptor metabotropic 7 is cis-regulated in the mouse brain and modulates alcohol drinking. *Genomics*, *90*, 690-702.

[83] Uhl, G. R., Drgon, T., Johnson, C., Li, C. Y., Contoreggi, C., Hess, J., Naiman, D. & Liu, Q. R. (2008). Molecular genetics of addiction and related heritable phenotypes: genome-wide association approaches identify "connectivity constellation" and drug target genes with pleiotropic effects. *Ann. N. Y. Acad. Sci.*, *1141*, 318-81.

[84] Mao, L., Guo, M., Jin, D., Xue, B. & Wang, J. Q. (2013). Group III metabotropic glutamate receptors and drug addiction. *Front. Med.*, *7*, 445-51.

[85] Blednov, Y. A., Walker, D., Osterndorf-Kahanek, E. & Harris, R. A. (2004). Mice lacking metabotropic glutamate receptor 4 do not show the motor stimulatory effect of ethanol. *Alcohol*, *34*, 251-9.

[86] Zaniewska, M., Przegalinski, E., Filip, M., Pilc, A. & Doller, D. (2014). Inhibitory actions of mGlu4 receptor ligands on cocaine-, but not nicotine-, induced sensitizing and conditioning locomotor responses in rats. *Pharmacol. Rep.*, *66*, 205-11.

[87] Nielsen, D. A., Ji, F., Yuferov, V., Ho, A., Chen, A., Levran, O., Ott, J. & Kreek, M. J. (2008). Genotype patterns that contribute to increased risk for or protection from developing heroin addiction. *Mol. Psychiatry*, *13*, 417-28.

[88] Olive, M. F., Cleva, R. M., Kalivas, P. W. & Malcolm, R. J. (2012). Glutamatergic medications for the treatment of drug and behavioral addictions. *Pharmacol. Biochem. Behav.*, *100*, 801-10.

[89] Reissner, K. J. & Kalivas, P. W. (2010). Using glutamate homeostasis as a target for treating addictive disorders. *Behav. Pharmacol.*, *21*, 514-22.

[90] Uys, J. D. & LaLumiere, R. T. (2008). Glutamate: the new frontier in pharmacotherapy for cocaine addiction. *CNS Neurol. Disord. Drug Targets*, *7*, 482-91.

[91] McClure, E. A., Gipson, C. D., Malcolm, R. J., Kalivas, P. W. & Gray, K. M. (2014). Potential role of N-acetylcysteine in the management of substance use disorders. *CNS Drugs*, *28*, 95-106.

[92] Asevedo, E., Mendes, A. C., Berk, M. & Brietzke, E. (2014). Systematic review of N-acetylcysteine in the treatment of addictions. *Rev. Bras. Psiquiatr.*, *36*, 168-75.

[93] Berk, M., Malhi, G. S., Gray, L. J. & Dean, O. M. (2013). The promise of N-acetylcysteine in neuropsychiatry. *Trends Pharmacol. Sci.*, *34*.

[94] Myers, K. M. & Carlezon, W. A. J. (2012). D-cycloserine effects on extinction of conditioned responses to drug-related cues. *Biol. Psychiatry*, *71*, 947-55.

[95] Kiefer, F., Kirsch, M., Bach, P., Hoffmann, S., Reinhard, I., Jorde, A., von der Goltz, C., Spanagel, R., Mann, K., Loeber, S. & Vollstadt-Klein, S. (2015). Effects of D-cycloserine on extinction of mesolimbic cue reactivity in alcoholism: a randomized placebo-controlled trial. *Psychopharmacology (Berl)*, in press.

[96] MacKillop, J., Few, L. R., Stojek, M. K., Murphy, C. M., Malutinok, S. F., Johnson, F. T., Hofmann, S. G., McGeary, J. E., Swift, R. M. & Monti, P. M. (2015). D-cycloserine to enhance extinction of cue-elicited craving for alcohol: a translational approach. *Transl. Psychiatry*, *5*, e544.

[97] Kamboj, S. K., Joye, A., Das, R. K., Gibson, A. J., Morgan, C. J. & Curran, H. V. (2012). Cue exposure and response prevention with heavy smokers: a laboratory-based randomised placebo-controlled trial examining the effects of D-cycloserine on cue reactivity and attentional bias. *Psychopharmacology (Berl).*, *221*, 273-84.

[98] Kamboj, S. K., Massey-Chase, R., Rodney, L., Das, R., Almahdi, B., Curran, H. V. & Morgan, C. J. (2011). Changes in cue reactivity and attentional bias following experimental cue exposure and response prevention: a laboratory study of the effects of D-cycloserine in heavy drinkers. *Psychopharmacology (Berl).*, *217*, 25-37.

[99] Price, K. L., Baker, N. L., McRae-Clark, A. L., Saladin, M. E., Desantis, S. M., Santa Ana, E. J. & Brady, K. T. (2013). A randomized, placebo-controlled laboratory study of the effects of D-cycloserine on craving in cocaine-dependent individuals. *Psychopharmacology (Berl).*, *226*, 739-46.

[100] Price, K. L., McRae-Clark, A. L., Saladin, M. E., Maria, M. M., DeSantis, S. M., Back, S. E. & Brady, K. T. (2009). D-cycloserine and cocaine cue reactivity: preliminary findings. *Am. J. Drug Alcohol Abuse*, *35*, 434-8.

[101] Prisciandaro, J. J., Myrick, H., Henderson, S., McRae-Clark, A. L., Santa Ana, E. J., Saladin, M. E. & Brady, K. T. (2013). Impact of DCS-facilitated cue exposure therapy on brain activation to cocaine cues in cocaine dependence. *Drug Alcohol Depend.*, *132*, 195-201.

[102] Gerrard, P. & Malcolm, R. (2007). Mechanisms of modafinil: a review of current research. *Neuropsychiatr. Dis. Treat.*, *3*, 349-64.

[103] Mahler, S. V., Hensley-Simon, M., Tahsili-Fahadan, P., Lalumiere, R. T., Thomas, C., Fallon, R. V., Kalivas, P. W. & Aston-Jones, G. (2014). Modafinil attenuates reinstatement of cocaine seeking: role for cystine-glutamate exchange and metabotropic glutamate receptors. *Addict. Biol.*, *19*, 49-60.

[104] Tahsili-Fahadan, P., Carr, G. V., Harris, G. C. & Aston-Jones, G. (2010). Modafinil blocks reinstatement of extinguished opiate-seeking in rats: mediation by a glutamate mechanism. *Neuropsychopharmacology*, *35*, 2203-10.

[105] Martinez-Raga, J., Knecht, C. & Cepeda, S. (2008). Modafinil: a useful medication for cocaine addiction? Review of the evidence from neuropharmacological, experimental and clinical studies. *Curr. Drug Abuse Rev.*, *1*, 213-21.

[106] Brady, K. T., Gray, K. M. & Tolliver, B. K. (2011). Cognitive enhancers in the treatment of substance use disorders: clinical evidence. Pharmacol. *Biochem. Behav.*, *99*, 285-94.

[107] Sofuoglu, M., Devito, E. E., Waters, A. J. & Carroll, K. M. (2013). Cognitive enhancement as a treatment for drug addictions. *Neuropharmacology*, *64*, 452-463.

[108] Blokhina, E. A., Kashkin, V. A., Zvartau, E. E., Danysz, W. & Bespalov, A. Y. (2005). Effects of nicotinic and NMDA receptor channel blockers on intravenous cocaine and nicotine self-administration in mice. *Eur. Neuropsychopharmacol.*, *15*, 219-25.

[109] Escher, T., Call, S. B., Blaha, C. D. & Mittleman, G. (2006). Behavioral effects of aminoadamantane class NMDA receptor antagonists on schedule-induced alcohol and self-administration of water in mice. *Psychopharmacology (Berl).*, *187*, 424-34.

[110] Aguilar, M. A., Manzanedo, C., Ribiero Do Couto, B., Rodriguez-Arias, M. & Minarro, J. (2009). Memantine blocks sensitization to the rewarding effects of morphine. *Brain Res.*, *1288*, 95-104.

[111] Krupitsky, E. M., Neznanova, O., Masalov, D., Burakov, A. M., Didenko, T., Romanova, T., Tsoy, M., Bespalov, A., Slavina, T. Y., Grinenko, A. A., Petrakis, I. L., Pittman, B., Gueorguieva, R., Zvartau, E. E. & Krystal, J. H. (2007). Effect of memantine on cue-induced alcohol craving in recovering alcohol-dependent patients. *Am. J. Psychiatry*, *164*, 519-23.

[112] Evans, S. M., Levin, F. R., Brooks, D. J. & Garawi, F. (2007). A pilot double-blind treatment trial of memantine for alcohol dependence. *Alcohol. Clin. Exp. Res.*, *31*, 775-82.

[113] Comer, S. D. & Sullivan, M. A. (2007). Memantine produces modest reductions in heroin-induced subjective responses in human research volunteers. *Psychopharmacology (Berl).*, *193*, 235-45.

[114] Jackson, A., Nesic, J., Groombridge, C., Clowry, O., Rusted, J. & Duka, T. (2009). Differential involvement of glutamatergic mechanisms in the cognitive and subjective effects of smoking. *Neuropsychopharmacology*, *34*, 257-265.

[115] Elkashef, A., Kahn, R., Yu, E., Iturriaga, E., Li, S. H., Anderson, A., Chiang, N., Ait-Daoud, N., Weiss, D., McSherry, F., Serpi, T., Rawson, R., Hrymoc, M., Weis, D., McCann, M., Pham, T., Stock, C., Dickinson, R., Campbell, J., Gorodetzky, C., Haning, W., Carlton, B., Mawhinney, J., Li, M. D. & Johnson, B. A. (2012). Topiramate for the treatment of methamphetamine addiction: a multi-center placebo-controlled trial. *Addiction*, *107*, 1297-306.

[116] Kenna, G. A., Lomastro, T. L., Schiesl, A., Leggio, L. & Swift, R. M. (2009). Review of topiramate: an antiepileptic for the treatment of alcohol dependence. *Curr. Drug Abuse Rev.*, *2*, 135-42.

[117] Minozzi, S., Cinquini, M., Amato, L., Davoli, M., Farrell, M. F., Pani, P. P. & Vecchi, S. (2015). Anticonvulsants for cocaine dependence. *Cochrane Database Syst. Rev.*, *4*, CD006754.

[118] Itil, T. M., Seaman, P. A., Huque, M., Mukhopadhyay, S., Blasucci, D., Tat, K. & Ciccone, P. E. (1978). The clinical and quantitative EEG effects and plasma levels of fenobam (McN-3377) in subjects with anxiety: an open rising dose tolerance and efficacy study. *Curr. Ther. Res.*, *24*, 708-724.

[119] Friedmann, C. T. H., Davis, L. J., Ciccone, P. E. & Rubin, R. T. (1980). Phase II double-blind controlled study of a new anxiolytic, fenobam (McN-3377) vs. placebo. *Curr. Ther. Res.*, *27*, 144-151.

[120] Pecknold, J. C., McClure, D. J. & Appeltauer, L. (1980). Fenobam in anxious outpatients. *Curr. Ther. Res.*, *27*, 119-123.

[121] Pecknold, J. C., McClure, D. J., Appeltauer, L., Wrzesinski, L. & Allan, T. (1982). Treatment of anxiety using fenobam (a nonbenzodiazepine) in a double-blind standard (diazepam) placebo-controlled study. *J. Clin. Psychopharmacol.*, *2*, 129-33.

[122] Goldberg, J., Gardos, G. & Cole, J. O. (1973). A controlled evaluation of pyrovalerone in chronically fatigued volunteers. *Int. Pharmacopsychiatry*, *8*, 60-9.

[123] Porter, R. H., Jaeschke, G., Spooren, W., Ballard, T., Buettelmann, B., Kolczewski, S., Peters, J. U., Prinssen, E., Wichmann, J., Vieira, E., Muehlemann, A., Gatti, S., Mutel, V. & Malherbe, P. (2005). Fenobam: a clinically validated non-benzodiazepine anxiolytic is a potent, selective and non-competitive mGlu5 receptor antagonist with inverse agonist activity. *J. Pharmacol. Exp. Ther.*, *315*, 711-721.

[124] Montana, M. C., Cavallone, L. F., Stubbert, K. K., Stefanescu, A. D., Kharasch, E. D. & Gereau, R. W. (2009). The metabotropic glutamate receptor subtype 5 antagonist fenobam is analgesic and has improved in vivo selectivity as compared to the prototypical antagonist 2-methyl-6-(phenylethynyl)-pyridine. *J. Pharmacol. Exp. Ther.*, *330*, 834-843.

[125] Berry-Kravis, E. M., Hessl, D., Coffey, S., Hervey, C., Schneider, A., Yuhas, J., Hutchison, J., Snape, M., Tranfaglia, M., Nguyen, D. V. & Hagerman, R. (2009). A pilot open-label single-dose trial of fenobam in adults with fragile X syndrome. *J. Med. Genet.*, *46*, 266-271.

[126] Mullard, A. (2015). Fragile X disappointments upset autism ambitions. *Nat. Rev. Drug Discov.*, *14*, 151-3.

[127] Keywood, C., Wakefield, M. & Tack, J. (2009). A proof of concept study evaluating the effect of ADX10059, a metabotropic glutamate receptor-5 negative allosteric modulator, on acid exposure and symptoms in gastro-esophageal reflux disease. *Gut*, *58*, 1192-9.

[128] Zerbib, F., Bruley des Varannes, S., Roman, S., Tutuian, R., Galmiche, J. P., Mion, F., Tack, J., Malfertheiner, P. & Keywood, C. (2011). Randomised clinical trial: effects of monotherapy with ADX10059, a mGluR5 inhibitor, on symptoms and reflux events in patients with gastro-oesophageal reflux disease. *Alim. Phar.*, *33*, 911-921.

[129] Jaeschke, G., Kolczewski, S., Spooren, W., Vieira, E., Bitter-Stoll, N., Boissin, P., Borroni, E., Buttelmann, B., Ceccarelli, S., Clemann, N., David, B., Funk, C., Guba,

W., Harrison, A., Hartung, T., Honer, M., Huwyler, J., Kuratli, M., Niederhauser, U., Pahler, A., Peters, J. U., Petersen, A., Prinssen, E., Ricci, A., Rueher, D., Rueher, M., Schneider, M., Spurr, P., Stoll, T., Tannler, D., Wichmann, J., Porter, R. H., Wettstein, J. G. & Lindemann, L. (2015). Metabotropic glutamate receptor 5 negative allosteric modulators: discovery of 2-chloro-4-[1-(4-fluorophenyl)-2,5-dimethyl-1H-imidazol-4-ylethynyl]pyridine (basimglurant, RO4917523), a promising novel medicine for psychiatric diseases. *J. Med. Chem.*, *58*, 1358-71.

[130] Lindemann, L., Porter, R. H., Scharf, S. H., Kuennecke, B., Bruns, A., von, K. M., Harrison, A. C., Paehler, A., Funk, C., Gloge, A., Schneider, M., Parrott, N. J., Polonchuk, L., Niederhauser, U., Morairty, S. R., Kilduff, T. S., Vieira, E., Kolczewski, S., Wichmann, J., Hartung, T., Honer, M., Borroni, E., Moreau, J. L., Prinssen, E., Spooren, W., Wettstein, J. G. & Jaeschke, G. (2015). Pharmacology of basimglurant (RO4917523, RG7090), a unique metabotropic glutamate receptor 5 negative allosteric modulator in clinical development for depression. *J. Pharmacol. Exp. Ther.*, *353*, 213-33.

[131] Gomez-Mancilla, B., Berry-Kravis, E., Hagerman, R., von Raison, F., Apostol, G., Ufer, M., Gasparini, F. & Jacquemont, S. (2014). Development of mavoglurant and its potential for the treatment of fragile X syndrome. *Expert Opin. Investig. Drugs*, *23*, 125-34.

[132] Dunayevich, E., Erickson, J., Levine, L., Landbloom, R., Schoepp, D. D. & Tollefson, G. D. (2008). Efficacy and tolerability of an mGlu2/3 agonist in the treatment of generalized anxiety disorder. *Neuropsychopharmacology*, *33*, 1603-1610.

[133] Bergink, V. & Westenberg, H. G. (2005). Metabotropic glutamate II receptor agonists in panic disorder: a double blind clinical trial with LY354740. *Int. Clin. Psychopharmacol.*, *20*, 291-3.

[134] Grillon, C., Cordova, J., Levine, L. R. & Morgan, C. A. 3rd. (2003). Anxiolytic effects of a novel group II metabotropic glutamate receptor agonist (LY354740) in the fear-potentiated startle paradigm in humans. *Psychopharmacology (Berl).*, *168*, 446-54.

[135] Kellner, M., Muhtz, C., Stark, K., Yassouridis, A., Arlt, J. & Wiedemann, K. (2005). Effects of a metabotropic $glutamate_{2/3}$ receptor agonist (LY544344/LY354740) on panic anxiety induced by cholecystokinin tetrapeptide in healthy humans: preliminary results. *Psychopharmacology (Berl).*, *179*, 310-5.

[136] Patil, S. T., Zhang, L., Martenyi, F., Lowe, S. L., Jackson, K. A., Andreev, B. V., Avedisova, A. S., Bardenstein, L. M., Gurovich, I. Y., Morozova, M. A., Mosolov, S. N., Neznanov, N. G., Reznik, A. M., Smulevich, A. B., Tochilov, V. A., Johnson, B. G., Monn, J. A. & Schoepp, D. D. (2007). Activation of mGlu2/3 receptors as a new approach to treat schizophrenia: a randomized Phase 2 clinical trial. *Nat. Med.*, *13*, 1102-1107.

[137] Kinon, B. J., Zhang, L., Millen, B. A., Osuntokun, O. O., Williams, J. E., Kollack-Walker, S., Jackson, K., Kryzhanovskaya, L. & Jarkova, N. (2011). A multicenter, inpatient, phase 2, double-blind, placebo-controlled dose-ranging study of LY2140023 monohydrate in patients with DSM-IV schizophrenia. *J. Clin. Psychopharm.*, *31*, 349-55.

[138] Stauffer, V. L., Millen, B. A., Andersen, S., Kinon, B. J., Lagrandeur, L., Lindenmayer, J. P. & Gomez, J. C. (2013). Pomaglumetad methionil: no significant difference as an

adjunctive treatment for patients with prominent negative symptoms of schizophrenia compared to placebo. *Schizophr. Res.*, *150*, 434-41.

[139] Liu, W., Downing, A. C., Munsie, L. M., Chen, P., Reed, M. R., Ruble, C. L., Landschulz, K. T., Kinon, B. J. & Nisenbaum, L. K. (2012). Pharmacogenetic analysis of the mGlu2/3 agonist LY2140023 monohydrate in the treatment of schizophrenia. *Pharmacogenomics J.*, *12*, 246-54.

In: Modulators of Glutamatergic Signaling
Editor: Zoran M. Pavlovic
ISBN: 978-1-63483-493-3

Chapter 4

MIGRAINE AND GLUTAMATE: MODULATORS OF GLUTAMATERGIC SIGNALING AS POTENTIAL TREATMENTS OF NEUROPSYCHIATRIC DISORDERS

***C. F. Gasparini*[1]*, R. A. Smith*[2] *and L. R. Griffiths*[2,*]**
[1]Menzies Health Institute Queensland, Griffith University Gold Coast, QLD, Australia
[2]Genomics Research Centre, Executive Director,
Institute of Health and Biomedical Innovation,
Queensland University of Technology, QLD, Australia

ABSTRACT

Migraine is a complex neurological disorder with a well-documented genetic basis. Migraine is a product of allelic variation in genes of neurological, vascular and hormonal origin interacting with environmental triggers. Presentation can include attacks of head pain with symptoms of nausea, emesis, photophobia, phonophobia, and occasionally, visual sensory disturbances, known as aura. Migraine pain is difficult to ignore, associated with a deep sense of malaise and manifests as a throbbing, pulsatile headache, localized to one side of the head that intensifies with physical activity and that can last from 4-72 hours. Migraine is diagnosed according to criteria developed by the International Headache Society (IHS) and is subdivided into two main types based on the occurrence of aura symptoms that may be present in the early stages of the headache: migraine with aura (MA) and migraine without aura (MO). The majority (about 70%) of migraineurs are diagnosed with the MO subtype whilst the remaining 30% experience MA accompanied by neurological symptoms that manifest as fully reversible, visual, sensory and/or dysphasic speech disturbances in conjunction with their headache. Glutamate is the primary excitatory neurotransmitter in the central nervous system (CNS) and over-excitation of glutamate receptors is regarded as a contributing factor, through various mechanisms, to the pathology of migraine. In this chapter we present an overview of the pathophysiology and co-morbidity of migraine with other psychiatric disorders and discuss the role of the glutamatergic system in migraine, its molecular components as

* Corresponding author: Professor Lyn Griffiths, Genomics Research Centre, Institute of Health and Biomedical Innovation, Queensland University of Technology, 60 Musk Ave, Kelvin Grove, QLD, 4059, Australia,.e-mail:lyn.griffiths@qut.edu.au.

potential drug targets, in addition to the current treatments and progress of modulators of glutamatergic signaling.

Keywords: glutamate, glutamatergic system, migraine treatment, migraine drugs, drug targeting, migraine with aura, migraine without aura

Migraine Historical Perspective

Migraine was researched and documented by a number of renowned physicians in the Roman era including Hippocrates, Arateus, Galen and Celsus, each making a valuable contribution to the knowledge-base of migraine pathophysiology. Hippocrates first described visual aura and the relief that occurs by purging in 400 BC [1]. Aretaeus of Cappadocia is typically mentioned as the "discoverer" of migraine based on historical texts he wrote on classification of headache into cephalgia, cephalea, and migraine in 100-200 AD [2]. Galen of Pergamon in approximately 200 AD introduced the Greek word *hemicrania*, a term that literally translates to mean "half" "skull" or "pain in half of the head" and from which the word migraine originates [3]. Celsus (215-300 AD) made the connection between migraine and certain external triggers such as wine, acidity, cold, or glaring brightness from fire or sun [4]. From these beginnings, knowledge of migraine pathophysiology has continued to evolve throughout history thanks to the cumulative contributions of many individuals.

Further historical milestones include the noted association of hormone fluctuations in women and precipitation of headache by Persian physician, Bakr Mohamed Ibn Zakariya Râzi (865–925AD). Also, the work of the Islamic scientist, Ibn Sina *(Avicenna*, 980–1037AD*)*, should receive mention for his tremendous contribution to medicine and migraine in the *Canon of Medicine*, a medical text that became one of the first scientific encyclopedias disseminated in Europe, within which he is thought to have hinted at a neurovascular theory of migraine. Later in the 16th century, Thomas Willis published a hypothesis linking 'megrim' to dilation of blood vessels and articulated a vascular theory of the disease [1]. Soon enough, this theory was challenged by Edward Lieving in 1873, who believed that migraine was more of a brain dysfunction caused by "nerve storms" and published a neurogenic theory of migraine [5]. Both theories have survived into the 20th century generating some debate about the origin and cause of headache and have produced a division of scientific thought regarding the applicability of theories of the pathogenesis of migraine.

References to migraine treatment are as old as 7,000 BC with remains of perforated Neolithic human skulls and historical cave paintings documenting the drastic steps taken by ancient people to combat migraine symptoms [6]. Prior to modern medicine, migraine pain was ascribed to the will of evil spirits or malevolent gods [7]. Remedies prescribed by ancient people included: "drilling a hole in the skull to free evil spirits, purging and bloodletting, applying a hot iron to the site of pain, inserting a clove of garlic through an incision in the temple, incantation and by remedies such as placing a clay crocodile on one's head to drive out the evil demons" [6]. Such archaic remedies were based on the supernatural including witchcraft, superstition and religious beliefs of the culture [8]. These primitive concepts are now obsolete and migraine is described as a complex neurological disorder influenced by an individual's genetic background. This chapter will briefly summarize the pathophysiology and co-morbidities of migraine and finally the available treatments and future therapies being developed targeting the glutamatergic system.

INTRODUCTION

Migraine is a disorder with a female preponderance (sex ratio of 3:1), which is correlated with variation in ovarian hormones and affects approximately 12% of studied Caucasian populations [9]. Migraine displays two main subtypes, termed migraine with aura or without aura (MA and MO respectively). The two forms are distinguished from each other based on the development of aura, a period of variable and diverse neurological symptoms that precede the headache phase of a migraine. Approximately 30% of migraineurs experience aura symptoms which include reversible numbness and tingling, speech disturbances, or visual disturbances, which may take the form of partial or complete loss of vision, as well as perception of unusual patterns or colours [10]. Diagnosis of migraine can be challenging as there are no apparent biochemical/genetic markers or evident anatomical pathology to form the basis of a diagnostic test. The diagnosis of migraine is thus not always straightforward, as it is based on subjective criteria, which considers previous family history and the exclusion of secondary causes and can be complicated by co-morbidities and overlapping symptoms with other neurological disorders.

Episodes of migraine unfold in four distinct phases: the prodrome (premotionary phase), aura (for MA attacks only), the headache phase (may be absent in MA), and finally the postdrome (resolution) [11]. Each stage may not be experienced by all migraineurs due to the wide spectrum of phenotypes in each patient. The episode begins with the prodrome phase which consists of warning symptoms that precede the headache phase of the migraine attack by several hours or days, and that act as a warning signal to the migraineur for an oncoming episode [11]. The second phase is the Aura that is defined by the International Headache Society (IHS) as the "complex of neurological symptoms that occur just prior to or at the onset of the migraine headache." Aura, as mentioned above, consists of a temporary visual, sensory and/or speech disturbance which is also experienced by some people with epilepsy before or after a seizure and lasts from 5 minutes to 1 hour [10]. The third phase consists of the headache phase which is the most common and follows the aura, if present. During this stage migraineurs experience a moderate to severe intensity attack that lasts 4-72 hours. The pain is often of a throbbing nature and is unilateral. Physical activity exacerbates the intensity and this is often accompanied by nausea and vomiting as well as possible photophobia and phonophobia. The migraine episode finally ends with the postdrome phase which can be thought of as a time of recovery.

A comprehensive understanding of each of the stages of a migraine attack is necessary to better understand the pathophysiology of migraine and identify acute and preventive therapies. Ongoing quantitative clinical observations, imaging studies, electrophysiological studies, and therapeutic clinical trials continue to provide important new information regarding how migraine starts and progresses. Migraine is believed to have a significant genetic aetiology, however only some genes, the sum of which comprise only a portion of trait heritability, have at this point in time been conclusively linked to the migraine subtypes MA and MO. Therefore, current research is actively pursuing the identification of unknown migraine susceptibility genes. Genetic studies have shown the risk of migraine to be significantly higher in first degree relatives of migraineurs compared to relatives of controls. Migraine families are continuously being investigated to determine the mode of genetic transmission of migraine and most studies have concluded in favour of a multifactorial inheritance pattern. Population based twin studies have confirmed a genetic influence on migraine ranging from 0.34 to 0.57, also indicating a significant environmental component to the disease [12, 13]. A major gene(s) initiating general migraine pathogenesis has not yet been identified and may be unlikely due to the heterogeneous make-up of the disorder. Investigation into the genetic basis of migraine has concentrated on neurological, vascular and hormonal pathways, with the most recent addition of mitochondrial genes. These studies have utilized a variety of populations, including out-bred cohorts, family pedigrees and genetic isolates. An awareness of the genetic links in migraine is important to inform treatment and management options.

The inequality in the gender ratio 3:1 of female migraine sufferers observed in epidemiological data is an interesting feature of this disorder that may be explained by hormones [14]. Migraine prevalence in women increases considerably post puberty, peaking at 35-45 years of age and declining post menopause [15, 16], indicating an involvement of hormones or hormonal imbalance in migraine manifestation. Indeed 7-8% of migraineurs are diagnosed into the Pure Menstrual Migraine subcategory of the disease and genetic studies have demonstrated a reproducible association of migraine with polymorphisms in the estrogen receptor gene *ESR1* in ethnically different populations [17]. Sex hormones are synthesized and metabolized in the brain and have potent effects on endogenous opioid, serotonergic, noradrenergic, β-adrenergic, dopaminergic, GABAergic systems and mitochondrial energy production [18]. However at present it is still unclear how ovarian hormones contribute to migraine, and an alternative hypothesis is that that there may be hidden genes on the X chromosome contributing to migraine phenotype due to the unusual majority of females suffering from migraine. The uneven gender distribution of migraine and the high proportion of male probands with affected first-degree relatives have led to the investigation of genomic factors on the X-chromosome. This may be mediated by gene dosing effects coupled with the fact that women have two copies of the X chromosome while men only have one X and one Y chromosome. Genetic studies investigating the X chromosome have identified three susceptibility loci at Xp22, Xq13 and Xq24-q28.

The study of a few single gene disorders that exhibit migraine with aura symptoms have helped expand the knowledge base regarding migraine and provide direction to the identification of genes in vascular and neurological families in common migraine [19]. These disorders include cerebral autosomal dominant arteriopathy with subcortical infarcts and leukoencephalopathy (CADASIL) syndrome, which presents with recurrent subcortical ischemic strokes and MA, and familial hemiplegic migraine (FHM). FHM is a rare sub-form of MA that follows strict autosomal dominant inheritance and was the first migraine syndrome to be linked to genetic mutations in ion channels, an observation which paved the way for further investigation of ion channel gene polymorphisms in migraine. Ion channel genes are particularly favourable targets in the study of migraine as they may promote Cortical Spreading Depression (CSD), a common feature during migraine attacks, and contribute to neuronal excitability. Mutations in three separate genes which are molecular components of ion transport and are involved in neuronal signaling have been identified as causing FHM and include: *CACNA1A*, *ATP1A2* and *SCNA1A* [20, 21]. The aim of current research is to develop a better understanding of the pathological mechanisms of migraine in these diseases and to study molecular gene-to-phenotype relationships to pathway-map gene defects and quantify the risk for developing migraine.

PATHOPHYSIOLOGY OF MIGRAINE

The focus of research on the pathophysiology of migraine has been to identify the source of the head pain and accompanying neurological symptoms. Theories to explain migraine pathogenesis have matured from the primary vascular hypothesis espoused by Wolff in the 1930s-1950s [22], which dominated much of history, to a currently modified neurovascular model which focuses on the interaction of vascular and neurological physiology. The neurovascular model implicates the trigeminovascular system as playing a critical role in the pathogenesis of migraine in combination with genetically disordered central neurotransmission [23]. The trigeminovascular system is composed of the trigeminal nerve, which is cranial nerve number V, the largest and most complex of the 12 cranial nerves, and the trigeminal innervation of the cranial vessels [24]. The trigeminal nerve consists of three large branches carrying fibres mediating sensory input via the ophthalmic, maxillary and mandibular divisions [24]. These nerve fibres innervate the skin, mucous membranes and sinuses of the face, the dura mater and large vessels of the cranial circulation and distribute to pain and temperature receptors in these tissues [25]. Activation of the trigeminal nerve in tandem with the release of inflammatory molecules including substance P, calcitonin gene related peptide (CGRP), and neurokinin A is believed to be the cause of pain sensations and other centrally mediated symptoms of migraine such as nausea, phonophobia and photophobia [26, 27]. These changes are associated with other neurological symptoms, such as Cortical Spreading Depression.

CORTICAL SPREADING DREPRESSION

In addition to implicating the trigeminovascular system, Cortical Spreading Depression (CSD), is a physiological event that has also been closely studied in migraine. CSD manifests as a slow wave of brain excitation characterized by abrupt shifts in ions across neuronal and glial cell membranes, particularly extracellular K^+ and glutamate, as well as fluxes in intracellular Na^+ and Ca^{2+} [28]. This slowly propagating wave is followed by sustained depression in activity and initiates in the occipital lobe, gradually spreading across the cortex. CSD is believed to be primarily responsible for the visual aura component of the headache [29].

Excess glutamate is thought to contribute to neuronal hyperexcitability, a mechanism that is implicated in both common and monogenic forms of MA [30]. Further evidence to support the CSD phenomenon as a migraine trigger comes from the study of mutations in the FHM genes; *CACNA1A* and *ATP1A2* in mouse models [31-35]. Known disease mutations in *CACNA1A* and *ATP1A2*, when introduced in knock-in mice, have been shown to cause excess release or slow uptake of glutamate from the synaptic cleft respectively, resulting in a reduced threshold for CSD induction and propagation [36]. CSD has also been shown to activate the trigeminovascular system in animal models and has been a target of anti-migraine drugs [28, 37].

COMORBIDITY OF MIGRAINE AND OTHER DISORDERS

Migraine in general is a disease that has a high occurrence of co-morbidities, as shown in numerous epidemiological studies. Co-morbidity is defined as the occurrence of two medical disorders in the same individual at the same time at a frequency greater than pure chance [9]. Firstly, migraine is associated with several psychiatric conditions, in particular depression, anxiety and panic disorders [38-43]. The mechanisms that lead to the co-morbidity of psychiatric disorders and migraine are not well understood and are a priority for future research, particularly as there is an overlap in symptoms between co-morbid disorders and in treatment options. The appearance of these co-morbidities can cause uncertainty in the diagnosis and complicate treatment of the disease in question leading to migraine being under-recognized, underdiagnosed, and undertreated, while potentially contributing to the originally identified psychological manifestation.

Migraine and epilepsy frequently co-occur together in the same individual more often than by coincidence. Migraine and epilepsy are both heterogeneous neurological disorders that share various risk factors, clinical features and both respond to the same drug therapy indicating a common pathophysiology [44]. Not surprisingly, anti-epileptic drugs, e.g., topiramate and valproate, which act by modulating neuronal function are often used as prophylactic migraine treatments and have demonstrated efficacy in double-blind trials, possibly because of an underlying shared neuronal hyperexcitability in both diseases [44-46]. Both disorders are chronic and defined by a characteristic temporary neurological dysfunction, sharing a number of common symptoms and similar stages of progression during an episode. These similarities often lead to one disorder being mistakenly diagnosed as the other [44]. Symptoms shared by these disorders include: post-event lethargy, impaired or loss of consciousness, visual disturbances, visual and hormonal triggering factors, vertigo, paraesthesias, hemiparesis and aphasia [45].

Studies have shown that both migraine and its co-morbid conditions are associated with poor health-related outcomes and the resultant decreased quality of life is intensified in an individual suffering from both migraine and another disease, such as depression [47]. Therefore recognition and classification of co-morbidities as separate entities, especially where they may be confused, is critical to minimizing adverse pharmacological events and reducing the burden of illness on health systems, families and work forces [48]. The study of co-morbidities can provide useful knowledge about disease pathophysiology and clues to uncovering causal pathways shared between migraine and its co-morbid disorders. Furthermore the characterization of co-morbidities can assist in identifying medications that are contraindicated and pathomechanisms in the co-morbid condition (for example triptans and ergotamines are contraindicated in cardiovascular diseases) [39]. This should reduce potential drug interactions from treating the two disorders simultaneously, but most importantly characterization of comorbidities can reveal novel pharmacological routes to pursue.

The Molecular Components of the Glutamatergic System As Potential Drug Targets

Glutamate is the most abundant excitatory neurotransmitter in the brain, with excitatory action that becomes excitotoxic when present at excessive extracellular concentrations. The molecular components of the glutamatergic system have been the focus of genetic studies in migraine due to accumulating evidence of perturbed glutamatergic homeostasis in plasma, platelets, urine and CSF of individuals afflicted with migraine (reviewed in Gasparini et al. 2015). The molecular components of the glutamatergic system which are responsible for glutamate synthesis, release and transport have not been thoroughly investigated at a genetic level in migraine. Single nucleotide polymorphisms (SNPs) in subunits of the ionotropic AMPA receptor have shown association with migraine, specifically two SNPs in *GRIA1* (rs548294 for MO, allelic P = 0.008, rs2195450 for MA, allelic P = 0.0005) and one SNP in *GRIA3* with MA (rs3761555 in females only, allelic P = 0.003) in a study by Formicola et al. 2010 [49]. In addition, a study by Maher et al. 2013 [50] in an Australian case-control cohort replicated the positive association in the *GRIA3* promoter polymorphism (rs3761555) initially identified by Formicola et al. in 2013 [49]. These are currently the only two genetic studies reporting a positive association in the gene constituents of the glutamatergic system with either MA or MO. In addition to these studies, however, there has been GWAS evidence which implicates glutamate in migraine indirectly. In a GWAS by Antilla et al. in 2010 one SNP, rs1835740, was positively associated with migraine. This SNP is located between the metadherin *MTDH* and *PGCP* genes which affect glutamate levels in the brain [51]. Additionally, there is evidence linking excess glutamate with CSD development in animal models, so pursuing modulators of the glutamatergic system for a potential role in migraine aetiology is a rational strategy [52]. Abnormal metabolism of the neurotransmitter glutamate is also implicated in a number of neurodegenerative and psychiatric diseases [53-55].

Receptors

Post-synaptic neurons synthesize and express distinct types of glutamatergic receptors on their cell surface to receive extracellular glutamate released by pre-synaptic neurons. Glutamatergic receptors are subdivided into two main types of membrane receptors, ionotropic and metabotropic. These two subtypes of receptors are distinguished by the mechanism by which glutamate activates them to transmit signal impulses in neurons. Ionotropic receptors (iGluRs) are gated to a ligand eg. Na^+, K^+, Ca^{2+}, or Cl^- and change conformation upon binding external glutamate, whereas metabotropic receptors employ G proteins to activate intracellular second messengers in their operation.

The glutamate receptor classification scheme recognizes three distinct types of ionotropic receptors: AMPA; NMDA; and kainate receptors. The name for each receptor subtype derives from the selective agonist that binds the receptor with highest affinity, specifically, α-amino-3-hydroxy-5-methyl-4-isoxazole propionate (AMPA) receptors,N-methyl-D-aspartate (NMDA) receptors, and ka (kainate) receptors (see Table 1) [56]. The ionotropic glutamate receptor (iGluR) subtypes, AMPA, NMDA and ka, are activated on a fast time scale (msec)

[57]. Each subunit possesses four transmembrane hydrophobic domains (designated TM1-4) within the central portion of the sequence.

Eight metabotropic receptor types have been classified into three groups (see Table 1). These receptors differ from ionotropic receptors in that they are coupled to G-proteins, which activate intracellular *biochemical cascades*, leading to the modification of other proteins, as for example ion channels. In contrast to the ionotropic receptors, metabotropic receptors operate on a time scale of several hundred milliseconds to seconds [57]. Metabotropic receptors are composed of seven transmembrane domains that span the cell membrane [58].

Table 1. The Biological and Gene Constituents of the Glutamatergic System

Enzymes	Gene	Chromosome (Human)
PDH, Pyruvate dehydrogenase	PDP1	8q22.1
PAG, Phosphate activated glutaminase	GLS2	12q13.3
mME, Mitochondrial malic enzyme	ME3	11cen-q22.3
GAD, Glutamic acid decarboxylase	GAD1	2q31
GS, Glutamine synthetase	GLUL	1q31
PC, Pyruvate carboxylase	PC	11q13.4-q13.5
cME, Cytosolic malic enzyme	ME1	6q12
AAT, Aspartate aminotransferase	GOT1	10q24.1-q25.1
GDH, Glutamate dehydrogenase	GLUD1	10q23.3
Ionotropic glutamate receptors (iGluRs)	**Gene**	**Chromosome (Human)**
AMPA	GRIA1	5q33
	GRIA2	4q32-33
	GRIA3	Xq25-26
	GRIA4	11q22-23
Kainate	GRIK1	21q21.1-22.1
	GRIK2	6q16.3-q21
	GRIK3	1p34-p33
	GRIK4	11q22.3
	GRIK5	19q34.3
NMDA	GRIN1	9q34.3
	GRIN2A	16p13.2
	GRIN2B	12p12
	GRIN2C	17q24-q25
	GRIN2D	19q13.1qter
	GRIN3A	9q31.1
	GRIN3B	19p13.3
Orphan	GRID1	10q
	GRID2	4q22
Metabotropic glutamate receptors (mGluRs)	**Gene**	**Chromosome (Human)**
mGluR1	GRM1	6q24
mGluR2	GRM2	3p21.2
mGluR3	GRM3	7q21.1-q21.2
mGluR4	GRM4	6p21.3
mGluR5	GRM5	11q14.3

Metabotropic glutamate receptors (mGluRs)	Gene	Chromosome (Human)
mGluR6	GRM6	5q35
mGluR7	GRM7	3p26-p25
mGluR8	GRM8	7q31.3-q32.1
Transporter Type	**Gene**	**Chromosome (Human)**
EAAT1	SLC1A3	5p13
EAAT2	SLC1A2	11p13
EAAT3	SLC1A1	9p24
EAAT4	SLC1A6	19p13.12
EAAT5	SLC1A7	1p32.3
VGLUT1	SCL17A7	9q13.33
VGLUT2	SCL17A6	11p14.3
VGLUT3	SCL17A8	12q23.1

Note: Some enzymes are composed from individual subunits that assemble to form larger multimeric complexes and therefore the same enzyme name will appear to have multiple chromosomal locations in the databases.

Transporters

Neurons must also be able to remove glutamate from the extracellular space to protect themselves from excitotoxicity damage. There are two distinct molecular transport systems that neurons use to protect themselves from the accumulation of extracellular glutamate and its neurotoxic effect, namely, plasma membrane transporters (EEATs) and vesicular transporters (VGLUTs) [59]. Five different 'high-affinity' glutamate transporters exist to inactivate glutamate, specifically: EAAT1; EAAT2; EAAT3; EAAT4 and; EAAT5 [60]. EAATs import extracellular glutamate into cells at the synapse by a high-affinity sodium-dependent process [61]. The re-uptake of excess glutamate at the synaptic cleft by EAATs is essential to recycling the amino acid pool and replenishing the supply of neurotransmitter in glutamatergic terminals [60]. Alongside the EAATs are the vesicular glutamate transporters (VGLUTs) whose job is to translocate glutamate into the lumen of pre-synaptic vesicles for intracellular storage [59]. Three VGLUT isoforms, VGLUT1, VGLUT2 and VGLUT3 have been molecularly identified and functionally characterized to date [62].

Evidence to Implicate the Glutamatergic System in Migraine

Regulation of glutamate synthesis, release, synaptic levels, and recycling are all key events that must be carefully coordinated for proper neuronal function and thus provide multiple targets for glutamatergic drug development. Due to the pervasive role of glutamate in the CNS deregulation of any of these events can potentially result in disease. Specifically, disease occurs when there are changes to the gating and conducting properties of a channel or if transport processes become defective. Perturbations in the functioning of these systems may manifest from mutations in the DNA sequence of the gene which can change the amino acid sequence and consequently the 3D structure and conformation of the protein, or alternatively mutations may affect the expression and regulation of the protein. Therefore it is important to characterize the genetic and expression profiles of the molecular components of the glutamatergic system as well as understand the signaling events from abnormal genes, or normal genetic variants to protein and clinical phenotype.

In a healthy state, glutamate is an essential excitatory amino acid neurotransmitter, however at excessive concentrations it begins to over-stimulate nerve cells, effectively killing them through excitotoxicity. Therefore, blocking the effects of excess glutamate which may contribute to neuronal hyperexcitability, a mechanism that is implicated in both common and monogenic forms of MA, could be a profitable approach that may help to reduce the transmission of afferent signals that turn into a migraine [30]. Regulating glutamate signaling using specific ion channel modulators that interact with glutamatergic channels in the membranes of neurons is the aim of the class of drug termed modulators. Effective modulators need to be specific in their mechanism of action in such a way that they do not interfere with other normal neuronal functions. Modulating nerve cells that disturb the normal pattern of brain circuit activity by blocking the cell surface receptors that bind glutamate may help reduce the state of hyperexcitability. Modulation of ionotropic and metabotropic glutamate receptors (NMDA, AMPA, ka) in migraine is a novel route of investigation that may contribute further understanding to the pathophysiology of migraine and its treatment. Importantly, many therapies for the acute treatment of migraine attacks have potent vasoconstrictor properties on the uterus and human coronary arteries in the heart and are therefore problematic for people that are at an increased risk of stroke and cardiovascular disease, which includes migraineurs themselves [63]. Therefore modulating the glutamatergic system presents a non-vasoactive alternative that could be advantageous to subgroups of migraineurs who only partially respond to mainstream products and do not gain adequate relief, or have elevated independent risk factors for cardiovascular disease.

Interest in investigating the glutamatergic system in migraine including its receptors and transporters has been fuelled by studies implicating glutamate in CSD [64, 65] development, manifesting in the aura as well as playing a role in activation of the trigeminovascular system and allodynia associated with migraine [66, 67] (see Table 2). Animal localization studies have also identified NMDA, AMPA, ka receptors in the trigeminal system [68, 69]. Inhibition of pain and prevention of central sensitization at the NMDA receptor by antagonists was a key discovery that has pinpointed these receptors as favorable targets for further research [70-72]. Animal and human models of pain have shown that the NMDA antagonists can prevent the induction of central sensitization and reverse existing central sensitization which are events implicated in migraine [70-72]. Localization experiments, preclinical pharmacologic studies and preclinical electrophysiologic experiments, all suggest excessive glutamate-mediated excitation in migraine. NMDA and non-NMDA receptors when blocked inhibit pain transmission to the trigeminal nucleus [69]. Glutamate receptors are reported to modulate capsaicin-induced c-fos expression in the trigeminal nucleus caudalis, which is an established model of migraine [73]. Overall it is important to understand which brain circuits may be disrupted in migraine and to then target these using NMDA receptor antagonists to restore signaling balance.

The three different ionotropic glutamate receptors have different roles in migraine and may act differently on migraine-related structures. Therefore it is imperative to untangle the exact role of each receptor subtype to understand the different consequences of targeting each different type, so that only specific signaling pathways are targeted. A prerequisite of clinically acceptable modulators is that they block excessive activation of receptors while preserving normal neurological function to minimize any side-effects. Despite some promising results with preclinical trials, the neurotoxic side-effects encountered in these compounds at the doses required to achieve the desired therapeutic benefits have not yet been outweighed and consequently many of these compounds have not yet continued to market. Structure-based drug design approaches as well as modeling and bioinformatics approaches may make it possible to circumvent some of these difficulties in the future.

Critical to the design and development of inhibitors is structural information about the key target protein. Integration of this information, including the key residues in the binding pocket and other allosteric sites that are involved in conferring affinity and cooperativity is vital to rational drug design of more potent and specific drugs. Detailed biophysical analysis of crystallographic and site-directed mutagenesis data of normal and mutated glutamate receptor channels is necessary to unravel the intricacies of channel function and reveal clues about ion channel gating and ion channel modulator mechanisms. This information will aid in designing more effective and specific therapeutic compounds based on the chemistry surrounding receptor-ligand interactions. Attention has also been given to the action of glutamate transporters in the disease process. Glutamate transport processes, which are carried out by EAATs and VGLUTs, are important in protecting neurons from excitotoxic neuronal damage caused by excessive glutamate signaling. Although there are no therapeutic compounds yet directed at EAATs or VGLUTs in preclinical development, future studies will be needed to test if these are also suitable targets to be modulated. There is genetic evidence in bipolar disorder pointing at EAATs and VGLUTs as being involved in disease, so their involvement in migraine is also possible [74].

CURRENT TREATMENTS AND MODULATORS OF GLUTAMATERGIC SIGNALING

There is no standard treatment for migraine, therapeutic intervention is based on physician prescription of individual medication by considering the level of pain intensity and disability, the presentation of key symptoms such as nausea and gastrointestinal disturbance, the presence of comorbid illness and the patient's previous history of response to headache medication, individual self-prescription, administration of alternative types of therapies and delivery systems (eg. oral, nasal, subcutaneous or rectal).

Acute medications are available over the counter or by prescription and taken as needed to stop the pain once it has begun. The majority of prescriptions filled for migraine are the triptans sumatriptan and maxalt, as these are often subsidised by government programmes and tend to be cheaper drugs than anticonvulsants. Modulators of the glutamatergic system that act through glutamate receptors that have undergone preclinical development for the treatment of migraine include LY293558, LY466195, BGG492, ADX10059 (see Table 3). These have been explored to varying degrees in preclinical and clinical studies and have served as prototypes for the development of modulators.

Table 2. Evidence to implicate the Glutamatergic System in Migraine

Type of Evidence	Example of Evidence
Genetic	Polymorphisms of GRIA genes in migraine aura patients FHM mutations induce CSD
Biochemical	Glutamatergic abnormalities in plasma, platelets, urine and CSF in migraine aura patients
Preclinical	Modulators of glutamate receptors in preclinical models and animal models of trigeminovascular activation support their involvement in migraine mechanisms
Physiologic	Animal localization studies have identified NMDA, AMPA, KA receptors in the trigeminal system Evidence that glutamate induces Cortical Spreading Depression Sensitization mechanisms

Table 3. Modulators of glutamatergic signaling in preclinical development

Study Drug	Receptor subtype
LY293558	AMPA/kainate
LY466195	kainate
BGG492	AMPA
ADX10059	mGluR5

Acute Antimigraine Treatments

Current treatment strategies for migraine involve two main approaches, symptomatic (acute) and preventive (prophylactic). Acute treatment aims to alleviate, abort, or reduce pain and accompanying symptoms of the headache after it has started. Although there are currently several mainstream pharmacotherapies for treating migraine symptoms or to reduce attacks, they affect a broad range of pharmacological targets and are therefore associated with side-effects and are not always efficacious for all migraine sufferers. Thus, there is strong impetus to develop more targeted treatments which are affordable and have reduced side-effects, as well as to understand the shared biology behind co-morbidities so as to gain insights into the pathophysiology underlying co-morbid disorders.

Many glutamatergic proteins remain uninvestigated and given the lack of available data on this topic, this is a mostly unexplored field of fundamental and applied research. Many questions concerning glutamate receptors and transporters are yet to be answered and in particular how their manipulation translates into a therapeutic effect. The ongoing research also needs to consider/evaluate optimizing properties such as the bio-availability, metabolism and receptor affinity of developed compounds, as well as considering the balance between clinical efficacy (clinical response rate 50%) and the side-effect profile.

LY293558 (Tezampanel)

LY293558, also known as tezampanel, is a non-selective AMPA/ka receptor antagonist that demonstrated efficacy for pain relief in allodynia, hyperalgesia, and lumbar monoradiculopathy [73]. The potential efficacy of LY293558 in migraine was also evaluated in a small proof of concept pilot study (n = 45) [75]. In this randomized triple-blind trial, 1.2mg/kg intravenous (IV) LY293558 was compared to 6 mg subcutaneous (SC) sumatriptan included as a positive control and placebo. Results indicated the drug LY293558 was significantly superior to placebo on all efficacy measures with a response rate of 69% for LY293558 versus 86% the active control sumatriptan and 25% for placebo [75]. The alluring feature of using LY293558 in clinical practice is that vasoconstrictive effects are eliminated, making it an option for treating patients with coronary artery risk factors. Generally, LY293558 was good in relieving symptoms of pain, nausea, sensitivity to light and sound, allodynia (sensitive/painful areas on the skin) with low incidence of adverse events [75, 76]. These preliminary results suggest that LY293558 could be useful in the treatment of migraine and should perhaps undergo further investigation.

LY466195

Although less studied, LY466195, is a competitive, selective antagonist of the GLUK5 receptor that has helped establish a role for this receptor in migraine. Expression of GLUK5 and GLUK2 receptors has been demonstrated in trigeminal neurons [68]. Additionally, LY466195 has been shown to competitively antagonize the GLUK5 receptor in a dural plasma protein extravasation (PPE) model of migraine [77]. This model studies activation of the trigeminal nerve and the inhibition of neurotransmitter release from the periphery [78]. PPE is an experimental model of migraine that refers to the leakage of plasma proteins from the dura (the outer meningeal layer) or 'extravasation.' Experimentally LY466195 was shown to block dural extravasation in Sprague-Dawley rats. LY466195 was also active in the c-Fos model where it blocked Fos protein expression in the nucleus caudalis. The upside of this compound LY466195 is that it does not cause vasoconstriction and this has been demonstrated in the rabbit saphenous vein.

BGG492

More recently a novel AMPA receptor antagonist and anticonvulsant, BGG492, has been developed by Novartis [79]. BGG492 is also in clinical development for treatment of epilepsy and tinnitus [80]. The efficacy, safety and tolerability of BGG492 was assessed in a randomized, double-blind Phase II study with 75 subjects with acute migraine attacks recruited and randomized to receive either BGG492 (250 mg), sumatriptan (100 mg) or placebo [79]. Overall, although improvement from severe or moderate to mild or no headache pain was reported, the proof-of-concept criterion for BGG492 was not met, requiring a 25% or greater improvement over placebo at two time points. Further studies must be done to further characterize this drug before it can be used effectively in the treatment of migraine.

ADX10059

ADX10059 is a negative allosteric modulator targeting the metabotropic glutamate receptor 5 (mGluR5) [81]. ADX10059 modulates the activity of the mGluR5 receptor by binding to an allosteric site different to the glutamate binding site, which is also known as the orthosteric agonist binding site [82]. ADX10059 is only effective when it binds to the allosteric site in the presence of the orthosteric ligand, glutamate. Hence ADX10059 is a negative modulator (NAM) that acts dynamically with the natural ligand glutamate on the receptor and when bound diminishes the intra-cellular signal generated by glutamate and therefore has inhibitory effects which are non-competitive [82]. ADX10059 was tested in a randomised clinical proof of concept study for acute treatment of migraine. In this study, ADX10059 met the primary efficacy objective, resulting in pain-free rates 16.1% with ADX10059 versus 4.5% with placebo ($p < 0.05$). Despite this positive result ADX10059 unfortunately has not progressed further at this stage due to liver enzyme changes and central nervous system adverse events which contraindicate its use. Nevertheless it is an example of a glutamatergic modulator that has shown through clinical trials to have the ability to reduce symptoms of acute migraine [83].

PROPHYLACTIC MIGRAINE TREATMENTS

Preventive treatment is used to decrease the frequency, intensity, and duration of attacks and includes identifying potential triggers. Therefore preventive treatment can be given even in the absence of a headache. The drugs lamotrigine, ketamine, topiramate, memantine, and Botulinum toxin type A are modulators of the glutamatergic system that have gained FDA approval and are on the market for the treatment of migraine as well as other conditions (see Table 4). They also have a robust trial history because they have also been tested in other neurologic diseases. They exhibit commonality, a concept in drug design whereby one drug can be used to treat more than one disease. These drugs offer opportunity for further study.

Table 4. FDA approved modulators of glutamatergic signaling

Drug	Activity
Lamotrigine	Blockage of voltage-gated Na^+ channels and N- and P/Q-type voltage-gated Ca^{2+} channels and inhibition of glutamate release
Ketamine	NMDA receptors
Topiramate	Blockage of voltage-dependent sodium channels, augmentation of GABA activity at some subtypes of the GABA-A receptors, antagonism of AMPA/kainate subtype of the glutamate receptor, and inhibition of the carbonic anhydrase enzyme, particularly isozymes II and IV
Memantine	NMDA receptors
Botulinum toxin type A	Reduces the release of glutamate

Anticonvulsants

There are various medicines that collectively are termed 'anticonvulsant drugs' or more accurately antiepileptic drugs (abbreviated "AEDs"), because they are typically used to treat epilepsy. Anticonvulsants as the name implies prevent convulsions and do this by modulating glutamate-mediated and/or the GABA-mediated systems by interacting with voltage-gated sodium (Na^+) and calcium (Ca^{2+}) channels [84]. The use of this class of drugs in the treatment of migraine is based on the rationale that migraine and epilepsy may share pathogenetic mechanisms. Anticonvulsant drugs can reduce the frequency of migraine attacks by 50% or more than with an inactive placebo [85]. Anticonvulsants such as topiramate, valproate, gapapenti, and lamotrigine which are typically used to treat epilepsy have also shown success in treating migraine, suggesting that the two disorders possibly share overlapping disease mechanisms. The use of anticonvulsant drugs for migraine prevention began with carbamazepine in 1970 [86]. The therapeutic potential of anticonvulsants in relieving migraine can be explained by their ability to decrease glutamate levels and enhance GABA and decrease abnormal excitability in the brain [87]. Despite a few reports describing efficacy of anticonvulsants flunarizine, sodium valproate, lamotrigine, verapamil and acetazolamide in FHM, anticonvulsants are generally considered second-line prophylactic agents and are used less often in the treatment of migraine. This is due to the fact that these drugs are more expensive and that the preferred first-line treatment are the triptans [88].

Lamotrigine

Lamotrigine works on glutamate signaling and was first used to treat epilepsy in 1994 in the US and since then its applicability in psychiatry has increased, especially as a treatment for bipolar disorder [89, 90]. Lamotrigine is a broad spectrum anticonvulsant whose mechanism of action is thought to be via blockage of voltage-gated Na^+ channels and N- and P/Q-type voltage-gated Ca^{2+} channels and inhibition of glutamate release [89, 91, 92]. The effect of lamotrigine in ameliorating aura has been demonstrated in a few studies. In a randomized double-blind trial of lamotrigine versus placebo in the prophylaxis of migraine with and without aura Steiner et al. treated 37 patients with lamotrigine and 40 with placebo for up to 3 months although only 53 patients reached the end of the full treatment period [93]. Following reports of side-effects of skin rash after an intial high dose of 200mg/day lamotrigine, a slow-dose escalation of 25mg/day for 2 weeks, 50mg/day for 2 weeks then 200mg/day was implemented [94]. The results of this trial did not demonstrate a statistically significant effect over placebo.

Lampl et al.1999 showed in a pilot study of 15 migraine with aura and only aura patients, that a dose of 25mg-100mg lamotrigine reduced aura frequency and duration [95]. The treatment period lasted 4 months and was followed by 3 months of post-treatment where participants ceased taking lamotrigine and recurring aura symptoms were monitored. During the post-treatment period it was observed that aura frequency and duration increased. Similarly in an open pilot study by D'Andrea et al. 13 of 21 patients affected by migraine with aura with a high frequency of attacks had their attacks completely eliminated. This study determined that lamotrigine should undergo further evaluation based on the observed reduction of migraine attacks [96].

In a small case study consisting of two patients with persistent aura like visual phenomena for months to years, symptoms disappeared after treatment with lamotrigine for a period of 2 months [97]. In another open study, 47 migraine patients with severe auras were treated with 100–150mg lamotrigine daily for 3 to 6 months [98]. The main outcome parameter was a >50% reduction in the mean frequency of migraine auras at 3 to 6 months of treatment, which the study achieved, obtaining greater than 75% reduction for 70% of the enrolled participants [98]. In a larger 3 year prospective study of 59 migraine with aura patients lamotrigine effectively reduced migraine aura and migraine attacks. Patients were treated with a dose of 25mg initially and increased as necessary until symptoms improved, but not exceeding a dose of 300mg lamotrigine daily [99]. This study found that 75% of the patients experienced an improvement in aura frequency and duration [99]. Participants took 5 months to reach the primary endpoint and there was no placebo. Skin rash was reported as the predominant side effect whilst on lamotrigine.

In a case study consisting of a small FHM2 family with a novel ATP1A2 mutation, prophylactic treatment with sodium valproate alone and in combination with lamotrigine reduced migraine occurrence almost completely [100]. In the father, when aged 38 years and his attack frequency occurred 3 times a week, 500mg of Na valproate daily reduced his attack frequency to a maximum of 1 per year for 8.5 years until publication of the study [100]. In his twin sons who like him suffered migraine from a young age, combination therapy of 1000mg of Na-valproate and 100mg lamotrigine per day stopped the attacks completely with no apparent side-effects [100]. This case study is a good example that highlights the important use of lamotrigine in treating migraine. In conclusion lamotrigine used either as a montherapy or in combination with other drugs may be useful for migraine patients particularly those with disturbing auras given the drug's ability to inhibit the neuronal release of glutamate and the fact that this is an event thought to be central to the genesis of migraine.

Ketamine

Ketamine, one of the most common NMDA receptor antagonists was first used in relation to migraine to treat aura symptoms. Ketamine is known to target the PCP site one of two available sites at the NMDA receptor. In a small study of 11 patients with familial hemiplegic migraine a trial dose of 25mg ketamine was administered as a nasal spray under medical supervision [101]. The aim of this study was to treat patients that had severe, disabling auras resulting from familial hemiplegic migraine. In five patients ketamine reproducibly reduced the severity and duration of the neurologic deficits, whereas in the remaining six patients no beneficial effect was seen. The six patients who failed to report a benefit experienced some side-effects including feelings of alienation and mild ataxia [101]. Due to the severity of aura symptoms and psychotropic side-effects of ketamine this study was conducted unblinded. This study concluded that NMDA receptor antagonists may be useful in the treatment of migraine aura.

Ketamine was shown to reduce aura severity but not duration in a randomized double-blind controlled trial of 25mg intranasal ketamine in 30 migraine patients with prolonged aura [102]. The aura can last from 5mins to 20mins and familial or sporadic hemiplegic migraine patients are the worst to be affected, though some experience extended periods of aura, having migraine with prolonged aura where the aura lasts longer than 1 hour and can cause disability for hours or days [10]. The reports of ketamine's rapid antidepressant effects coupled with preclinical reports from a variety of other agents acting at the NMDA receptor have spurred increasing interest in the development of NMDA modulators as novel therapeutics in mood disorders. Although ketamine has been tested and evaluated for migraine prophylaxis it is generally rarely prescribed for this purpose. This is because ketamine is a controlled substance, with a relatively high potential for addiction and abuse that can easily, if not regulated, lead to moderate/low physical or psychological dependence. Nonetheless due to its modulatory role on the glutamatergic system ketamine has been useful to advance knowledge of the mechanism by which it may modulate NMDA receptors.

Topiramate

Topiramate was first used as an anticonvulsant therapy for epilepsy but is now also used to treat migraine in certain cases. There is an overabundance of literature in regards to topiramate and it is the most promising compound that has progressed the farthest along the experimental drug pipeline supported by clinical trials. Topiramate is a derivative of the naturally occurring monosaccharide D-fructose and although the exact mechanism of action is not completely understood, topiramate's efficacy in migraine prevention is believed to be a result of its neuro-stabilizing properties. The neuro-stabilizing properties of topiramate include blockage of voltage-dependent sodium channels, an augmentation of GABA activity at some subtypes of the GABA-A receptors, antagonism of AMPA/kainate subtype of the glutamate receptor, and inhibition of the carbonic anhydrase enzyme, particularly isozymes II and IV [103]. These properties may act to reduce neuronal hyperexcitability and cortical spreading depression [104]. Topiramate is known to inhibit the excitatory activity of glutamate by acting at the receptor subtype for kainate/AMPA [103]. The effectiveness of topiramate has been rigorously established in several small randomized controlled trials and two large multicentre randomized controlled trials as well as various open-label observational studies [104-106].

Memantine

Memantine is a derivative of adamantine which specifically targets the glutamatergic NMDA receptors. Memantine is thought to reduce brain activity by binding and blocking the activity of NMDA receptors and thereby decreasing the effects of glutamate and protecting nerve cells from excessive stimulation by glutamate [107]. The drug is clinically effective because of moderate-affinity non-competitive antagonism with a fast off-rate that as a result spares normal neurotransmission which is paramount to normal brain activity [108]. Memantine is well-tolerated clinically and has many therapeutic applications ranging from Alzheimer's disease to dementia, Parkinson's and Huntington's diseases because it can interact with a variety of ligand-gated ion channels [109]. Memantine was first marketed and approved for use for the treatment of moderate to severe Alzheimer's disease in the United States in 2003. Although memantine's intended use is for the treatment of Alzheimer's disease, it can be prescribed as an off-label prophylactic therapy for migraine [110-112].

Memantine is believed to work by binding at or near the Mg^{2+} site in the NMDA receptor and helps to block over-activation of the receptor only under excitotoxic pathological conditions [113]. Activation of the NMDA receptor occurs when two ligands glutamate and glycine or D-serine both are bound to the NMDA receptor. When the NMDA receptor is activated it allows an influx of positively charged ions. In addition, endogenous Mg^{2+} helps to regulate the flow of current through the NMDA receptor in a voltage-dependent manner rendering the channel permeable to Ca^{2+} which acts as a messenger for intracellular processes via the activation of Ca^{2+}-dependent protein kinases [114, 115].

In a clinical trial evaluating the efficacy of memantine on migraine, 54 subjects were treated beginning at a dose of 5mg/day, increasing if needed by 5mg/week up to 10mg twice a day [116, 117]. The patient group consisted of 20 MA and 40 MO migraineurs with an average age of 49 whose previous treatment approaches were unsuccessful. The data showed that 67% of participants receiving memantine reported a greater than 50% reduction in estimated monthly headache frequency after the 2 month trial [116]. Although side-effects were reported in approximately 10% of patients and resulted in discontinuation of the medication, overall the results of this trial indicate that the use of memantine lead to a significant reduction in average headache frequency, and improved function and may work for some migraineurs. This trial was limited by the fact that it was retrospective, had no control group or blinding.

In a different clinical study, memantine was investigated for refractory migraine [118]. Refractory migraine is also known as chronic migraine which is a headache that continues persistently for at least 15 days a month, with at least eight days a month on which the headache and associated symptoms meet diagnostic criteria for migraine and do not respond to standard treatments [119]. Twenty-three of 38 recruited subjects completed the 3 month study and received between 10mg and 20mg of memantine [118]. The average number of headache days per month was reduced from 21.8 to 16.1 headache days per month after 3 months ($P < .01$) in patients already under prevention with one or more drugs. Noteworthy was that side-effects were present in 37.5% of the patients and 5.5% of patients dropped out of the study [118]. This trial was limited by the lack of a placebo and the fact that migraine patients continued to take preventive medications during the trial and there was no blinding.

Memantine has also been evaluated for prophylaxis in a double-blind, randomized, crossover clinical trial of chronic tension-type headache patients [120]. Of 40 patients who enrolled in the trial, 29 continued therapy and completed the study and received 20-40mg/day memantine or placebo for 10 weeks. The primary efficacy end-point for the study population was taken as the area under the headache curve (duration x intensity), this measure did not differ between memantine and placebo [120]. Memantine was, however, shown to modestly reduce headache intensity [120].

Botulinum Toxin Type A

Botulinum toxin type A (BoNTA) is a medical product composed of tiny quantities of a highly purified and potent neurotoxin (botulinum toxin type A) protein refined from the bacterium, *Clostridium botulinum* [121]. BoNTA is only accessible with a prescription and although it is used in different applications and new uses continue to emerge it is typically injected into a specific muscle or gland to achieve its therapeutic benefit. BoNTA is thought to cause muscle relaxation by blocking the vesicular release of the neurotransmitter acetylcholine at the neuromuscular junction [122]. Due to its muscle paralysing effect BoNTA has found wide application in the treatment of muscular conditions and is well known for its use in the cosmetic industry to reduce wrinkles [123, 124]. BoNTA was first approved by the FDA in 2010 and manufactured by the pharmaceutical company Allergan to prevent headache in patients with chronic migraine by specialist neurologists [125]. Patients with chronic migraine are defined as having headaches on at least 15 days a month, with at least eight days a month on which their headaches and associated symptoms meet diagnostic criteria for chronic migraine [119, 126]. Currently BoNTA is approved for the treatment of chronic migraine and is administered as a fixed site/fixed dose injection in the temporalis muscle for the relief of chronic head pain in eligible patients [127]. Bittencourt da Silva et al. designed an experimental model of pain consisting of a topical capsaicin patch (8%) plus mild heat to assay the effect of BoNTA on evoked pain and vasomotor responses (temperature and blood flow) and glutamate release in the skin [128]. This study demonstrated that BoNTA, reduced the release of glutamate in the skin by a mechanism not fully understood which may involve other undiscovered neuropeptides/neurotransmitters which are involved in pain induction or transmission contributing to BoNTA's effect. These findings have also been observed in *in vitro* culture systems and *in vivo* animal studies [129-131]. Research on BoNTA for migraine treatment and to investigate the effect and mechanism of BoNTA on evoked pain is ongoing [132].

CONCLUSION

Conclusion and Future Perspectives on Glutamate Modulating Drugs

Migraine is a debilitating neurological disorder with an undetermined cause resulting in painful attacks of headache that have been linked to neuronal hyperexcitability. Genetic, biochemical and pharmacological studies have identified positive associations within a few key genetic variants that in conjunction with biochemical findings support perturbed glutamate homeostasis in peripheral tissues of migraineurs and altered function of the glutamatergic system in migraineurs as being involved in the disease process. More detailed and precise characterization of the molecular components of the glutamatergic system is warranted as an initial starting point to inform future structure-based drug design efforts and to also elucidate the mechanism by which these drugs work, so as to develop novel compounds that are more potent and specific. Although great strides in the pharmacology of glutamatergic receptors have occurred, some stumbling blocks to translating prototypical agonists into the clinical setting still remain. Mainly these include: i) off target actions at other receptor sites; ii) a lack of on target efficacy in the disease setting; iii) on target adverse effects. The question remains whether we will be able to tilt the balance in favour of a therapeutic effect while we are still wrestling with side-effects and unwanted off target actions. Nevertheless, glutamate receptors remain interesting targets to investigate for drug development in relation to migraine. More work, however, is necessary to fill some gaps in our understanding of key processes and their mechanism of action. In essence, more complete understanding of the complex labyrinth of the glutamatergic system and its modulators in migraine along with pharmaco-genomic discoveries is expected to come in the next few years as further experiments, projects and clinical trials are conducted.

REFERENCES

[1] Abokrysha, N. (2009). Ibn Sina (Avicenna) on pathogenesis of migraine compared with the recent theories. *Headache.*, *49*(6), 923-927.

[2] Koehler, P. J. & van de Wiel, T. W. (2001). Aretaeus on migraine and headache. *J Hist Neurosci.*, *10*(3), 253-261.

[3] Afridi, S. & Kaube, H., (2003). Prophylactic Therapy for Migraine. *Curr Treat Options Neurol.*, *5*(6), 431-440.

[4] Silberstein, S. D., Stiles, M. A. & Young, W. B. editors (2005) Atlas of migraine and other headaches. 2nd ed. London: Taylor and Francis.

[5] Eadie, M. J. (2005). The pathogenesis of migraine - 17th to early 20th Century understandings. *Journal of Clinical Neuroscience.*, *12*(4), 383-388.

[6] Villalon, C. M., Centurion, D., Valdivia, L. F., de Vries, P. & Saxena, P. R. (2003). Migraine: pathophysiology, pharmacology, treatment and future trends. *Current Vascular Pharmacology.*, *1*(1), 71-84.

[7] Takano, T. & Nedergaard, M. (2009). Deciphering migraine. *J Clin Invest.*, *119*(1), 16-19.

[8] Rapoport, A. & Edmeads, J. (2000). Migraine - The evolution of our knowledge. *Archives of Neurology.*, *57*(8), 1221-1223.

[9] Lipton, R. B., Bigal, M. E., Diamond, M., Freitag, F. & Reed, M. L. et al. (2007). Migraine prevalence, disease burden, and the need for preventive therapy. *Neurology.*, *68*(5), 343-349.

[10] Viana, M., Sprenger, T., Andelova, M. & Goadsby, P. J. (2013). The typical duration of migraine aura: a systematic review. *Cephalalgia.*, *33*(7), 483-490.

[11] Charles, A. (2013). The evolution of a migraine attack - a review of recent evidence. *Headache.*, *53*(2), 413-419.

[12] Mulder, E. J., Van Baal, C., Gaist, D., Kallela, M. & Kaprio, J. et al. (2003). Genetic and environmental influences on migraine: a twin study across six countries. *Twin Research.*, *6*(5), 422-431.

[13] Svensson, D. A., Larsson, B., Waldenlind, E. & Pedersen, N. L. (2003). Shared Rearing Environment in Migraine: Results From Twins Reared Apart and Twins Reared Together. *Headache: The Journal of Head and Face Pain.*, *43*(3), 235-244.

[14] Merikangas, K. R. (2013). Contributions of Epidemiology to Our Understanding of Migraine. *Headache.*, *53*(2), 230-246.

[15] Jensen, R. & Stovner, L. J. (2008). Epidemiology and comorbidity of headache. *Lancet Neurol.*, *7*(4), 354-361.

[16] Stewart, W. F., Lipton, R. B., Celentano, D. D. & Reed, M. L. (1992). Prevalence of migraine headache in the United States. Relation to age, income, race, and other sociodemographic factors. *JAMA.*, *267*(1), 64-69.

[17] Russell, M. B. (2010). Genetics of menstrual migraine: the epidemiological evidence. *Curr Pain Headache Rep.*, *14*(5), 385-388.

[18] Shuster, L. T., Faubion, S. S., Sood, R. & Casey, P. M. (2011). Hormonal manipulation strategies in the management of menstrual migraine and other hormonally related headaches. *Current Neurology and Neuroscience Reports.*, *11*(2), 131-138.

[19] Liem, M. K., Oberstein, S. A., van der Grond, J., Ferrari, M. D. & Haan, J. (2010). CADASIL and migraine: A narrative review. *Cephalalgia.*, *30*(11), 1284-1289.

[20] de Vries, B., Frants, R. R., Ferrari, M. D. & van den Maagdenberg, A. M. (2009). Molecular genetics of migraine. *Hum Genet.*, *126*(1), 115-132.

[21] van den Maagdenberg, A. M., Pietrobon, D., Pizzorusso, T., Kaja, S., Broos, L. A., et al. (2004). A Cacna1a knockin migraine mouse model with increased susceptibility to cortical spreading depression. *Neuron.*, *41*(5), 701-710.

[22] Wolff, H.G. (2001) Wolff's Headache and Other Head Pain. New York: Oxford University Press (US).

[23] Welch, K. M., D'Andrea, G., Tepley, N., Barkley, G. & Ramadan, N. M. (1990). The concept of migraine as a state of central neuronal hyperexcitability. *Neurol Clin.*, *8*(4), 817-828.

[24] Sanders, R. D. (2010). The Trigeminal (V) and Facial (VII) Cranial Nerves: Head and Face Sensation and Movement. *Psychiatry (Edgmont (Pa. : Township)).*, *7*(1), 13-16.

[25] May, A. & Goadsby, P. J. (1999). The trigeminovascular system in humans: Pathophysiologic implications for primary headache syndromes of the neural influences on the cerebral circulation. *Journal of Cerebral Blood Flow and Metabolism.*, *19*(2), 115-127.

[26] Goadsby, P. J., Charbit, A. R., Andreou, A. P., Akerman, S. & Holland, P. R. (2009). Neurobiology of migraine. *Neuroscience.*, *161*(2), 327-341.

[27] Goadsby, P. J., Edvinsson, L. & Ekman, R. (1990). Vasoactive peptide release in the extracerebral circulation of humans during migraine headache. *Ann Neurol.*, *28*(2), 183-187.

[28] Charles, A. C. & Baca, S. M. (2013). Cortical spreading depression and migraine. *Nat Rev Neurol.*, *9*(11), 637-644.

[29] Costa, C., Tozzi, A., Rainero, I., Cupini, L. M., Calabresi, P., et al. (2013). Cortical spreading depression as a target for anti-migraine agents. *Journal of Headache and Pain.*, 14.

[30] Gorji, A., Scheller, D., Straub, H., Tegtmeier, F. & Kohling, R., et al. (2001). Spreading depression in human neocortical slices. *Brain Res.*, *906*(1-2), 74-83.

[31] van den Maagdenberg, A. M. J. M., Pietrobon, D., Pizzorusso, T., Kaja, S., Broos, L. A. M., et al. (2004). A Cacna1a Knockin Migraine Mouse Model with Increased Susceptibility to Cortical Spreading Depression. *Neuron.*, *41*(5), 701-710.

[32] Tottene, A., Conti, R., Fabbro, A., Vecchia, D., Shapovalova, M., et al. (2009). Enhanced Excitatory Transmission at Cortical Synapses as the Basis for Facilitated Spreading Depression in Ca(v)2.1 Knockin Migraine Mice. *Neuron.*, *61*(5), 762-773.

[33] Kaja, S., Van de Ven, R. C. G., Broos, L. A. M., Frants, R. R., Ferrari, M. D. et al., (2010). Severe and Progressive Neurotransmitter Release Aberrations in Familial Hemiplegic Migraine Type 1 Cacna1a S218L Knock-in Mice. *Journal of Neurophysiology.*, *104*(3), 1445-1455.

[34] van den Maagdenberg, A. M. J. M., Pizzorusso, T., Kaja, S., Terpolilli, N., Shapovalova, M., et al. (2010). High Cortical Spreading Depression Susceptibility and

Migraine-Associated Symptoms in Ca(V)2.1 S218L Mice. *Annals of Neurology.*, *67*(1), 85-98.

[35] Leo, L., Gherardini, L., Barone, V., De Fusco, M., Pietrobon, D., et al. (2011). Increased susceptibility to cortical spreading depression in the mouse model of familial hemiplegic migraine type 2. *PLoS Genetics.*, *7*(6), e1002129.

[36] Pietrobon, D. (2005). Function and dysfunction of synaptic calcium channels: insights from mouse models. *Current Opinion in Neurobiology.*, *15*(3), 257-265.

[37] Bolay, H., Reuter, U., Dunn, A. K., Huang, Z., Boas, D. A. et al. (2002). Intrinsic brain activity triggers trigeminal meningeal afferents in a migraine model. *Nature Medicine.*, *8*(2), 136-142.

[38] Hamelsky, S. W. & Lipton, R. B. (2006). Psychiatric comorbidity of migraine. *Headache.*, *46*(9), 1327-1333.

[39] Lal, V. & Singla, M. (2010). Migraine comorbidities--a discussion. *J Assoc Physicians India.*, *58* Suppl(18-20.

[40] Bag, B., Hacihasanoglu, R. & Tufekci, F. G. (2005). Examination of anxiety, hostility and psychiatric disorders in patients with migraine and tension-type headache. *International Journal of Clinical Practice.*, *59*(5), 515-521.

[41] Mongini, F., Rota, E., Deregibus, A., Ferrero, L., Migliaretti, G. et al., (2006). Accompanying symptoms and psychiatric comorbidity in migraine and tension-type headache patients. *Journal of Psychosomatic Research.*, *61*(4), 447-451.

[42] Oedegaard, K. J., Neckelmann, D., Mykletun, A., Dahl, A. A., Zwart, J. A. et al. (2006). Migraine with and without aura: association with depression and anxiety disorder in a population-based study. The HUNT Study. *Cephalalgia.*, *26*(1), 1-6.

[43] Zwart, J. A., Dyb, G., Hagen, K., Odegard, K. J., Dahl, A. A. et al. (2003). Depression and anxiety disorders associated with headache frequency. The Nord-Trondelag Health Study. *Eur J Neurol.*, *10*(2), 147-152.

[44] Bigal, M. E., Lipton, R. B., Cohen, J. & Silberstein, S. D. (2003). Epilepsy and migraine. *Epilepsy Behav.*, *4* Suppl 2(S13-24.

[45] De Simone, R., Ranieri, A., Marano, E., Beneduce, L., Ripa, P. et al. (2007). Migraine and epilepsy: clinical and pathophysiological relations. *Neurol Sci.*, *28* Suppl 2(S150-155.

[46] Haan, J., Terwindt, G. M., van den Maagdenberg, A. M., Stam, A. H. & Ferrari, M. D. (2008). A review of the genetic relation between migraine and epilepsy. *Cephalalgia.*, *28*(2), 105-113.

[47] Jette, N., Patten, S., Williams, J., Becker, W. & Wiebe, S. (2008). Comorbidity of migraine and psychiatric disorders--a national population-based study. *Headache.*, *48*(4), 501-516.

[48] Silberstein, S. D., Dodick, D., Freitag, F., Pearlman, S. H., Hahn, S. R. et al. (2007). Pharmacological approaches to managing migraine and associated comorbidities--clinical considerations for monotherapy versus polytherapy. *Headache.*, *47*(4), 585-599.

[49] Formicola, D., Aloia, A., Sampaolo, S., Farina, O., Diodato, D. et al. (2010). Common variants in the regulative regions of GRIA1 and GRIA3 receptor genes are associated with migraine susceptibility. *Bmc Medical Genetics.*, *11*(103-114.

[50] Maher, B. H., Lea, R., Follett, J., Cox, H. C. & Fernandez, F. et al. (2013). Association of a GRIA3 gene polymorphism with migraine in an Australian case-control cohort. *Headache: The Journal of Head and Face Pain.*, *53*(8), 1245-1249.

[51] Anttila, V., Stefansson, H., Kallela, M., Todt, U., Terwindt, G. M. et al. (2010). Genome-wide association study of migraine implicates a common susceptibility variant on 8q22.1. *Nature Genetics.*, *42*(10), 869-873.

[52] Fifkova E. & van Harreveld, A. (1974). Glutamate and spreading depression. *Journal of Neurobiology.*, *5*(5), 469-473.

[53] Javitt, D. C. (2004). Glutamate as a therapeutic target in psychiatric disorders. *Mol Psychiatry.*, *9*(11), 984-997, 979.

[54] Sanacora, G., Zarate, C. A., Krystal, J. H. & Manji, H. K. (2008). Targeting the glutamatergic system to develop novel, improved therapeutics for mood disorders. *Nat Rev Drug Discov.*, *7*(5), 426-437.

[55] Noetzel, M. J., Jones, C. K. & Conn, P. J. (2012). Emerging approaches for treatment of schizophrenia: modulation of glutamatergic signaling. *Discov Med.*, *14*(78), 335-343.

[56] Vikelis, M. & Mitsikostas, D. D. (2007). The role of glutamate and its receptors in migraine. *CNS Neurol Disord Drug Targets.*, *6*(4), 251-257.

[57] Gasic, G. P. & Hollmann, M. (1992). Molecular neurobiology of glutamate receptors. *Annual Review of Physiology.*, *54*(507-536.

[58] Julio-Pieper, M., Flor, P. J., Dinan, T. G. & Cryan, J. F. (2011). Exciting times beyond the brain: metabotropic glutamate receptors in peripheral and non-neural tissues. *Pharmacol Rev. 63*(1), 35-58.

[59] Liguz-Lecznar, M. & Skangiel-Kramska, J. (2007). Vesicular glutamate transporters (VGLUTs): the three musketeers of glutamatergic system. *Acta Neurobiol Exp (Wars).*, *67*(3), 207-218.

[60] Seal, R. P. & Amara, S. G. (1999). Excitatory amino acid transporters: a family in flux. *Annu Rev Pharmacol Toxicol.*, *39*(431-456.

[61] Tremolizzo, L., DiFrancesco, J. C., Rodriguez-Menendez, V., Sirtori, E., Longoni, M. et al., (2006). Human platelets express the synaptic markers VGLUT1 and 2 and release glutamate following aggregation. *Neuroscience Letters.*, *404*(3), 262-265.

[62] Benarroch, E. E. (2010). Glutamate transporters: diversity, function, and involvement in neurologic disease. *Neurology.*, *74*(3), 259-264.

[63] Tfelt-Hansen, P., Saxena, P. R., Dahlof, C., Pascual, J., Lainez, M. et al. (2000). Ergotamine in the acute treatment of migraine: a review and European consensus. *Brain.*, *123* (Pt 1)(9-18.

[64] D'Andrea, G. & Leon, A. (2010). Pathogenesis of migraine: from neurotransmitters to neuromodulators and beyond. *Neurol Sci.*, *31* Suppl 1(S1-7.

[65] Charles, A. & Brennan, K. (2009). Cortical spreading depression-new insights and persistent questions. *Cephalalgia.*, *29*(10), 1115-1124.

[66] Ramadan, N. M. (2003). The link between glutamate and migraine. *CNS Spectr.*, *8*(6), 446-449.

[67] Kaikai, M. A. & Howe, R. (1991). Glutamate-Immunoreactivity in the trigeminal and dorsal-root ganglia, and intraspinal neurons and fibers in the doral horn of the rat *Histochemical Journal.*, *23*(4), 171-179.

[68] Sahara, Y., Noro, N., Iida, Y., Soma, K. & Nakamura, Y. (1997). Glutamate receptor subunits GluR5 and KA-2 are coexpressed in rat trigeminal ganglion neurons. *J Neurosci.*, *17*(17), 6611-6620.

[69] Storer, R. J. & Goadsby, P. J. (1999). Trigeminovascular nociceptive transmission involves N-methyl-D-aspartate and non-N-methyl-D-aspartate glutamate receptors. *Neuroscience.*, *90*(4), 1371-1376.
[70] Block, F. & Habermeyer, B. (2003). Glutamate antagonists for treatment of neuropathic pain. *Schmerz.*, *17*(4), 261-267.
[71] Hewitt, D. J. (2000). The use of NMDA-receptor antagonists in the treatment of chronic pain. *Clinical Journal of Pain.*, *16*(2), S73-S79.
[72] Latremoliere, A. & Woolf, C. J. (2009). Central sensitization: a generator of pain hypersensitivity by central neural plasticity. *J Pain.*, *10*(9), 895-926.
[73] Sang, C. N., Hostetter, M. P., Gracely, R. H., Chappell, A. S., Schoepp, D. D. et al. (1998). AMPA/kainate antagonist LY293558 reduces capsaicin-evoked hyperalgesia but not pain in normal skin in humans. *Anesthesiology.*, *89*(5), 1060-1067.
[74] Machado-Vieira, R., Ibrahim, L., Henter, I. D. & Zarate, C. A. (2012). Novel glutamatergic agents for major depressive disorder and bipolar disorder. *Pharmacology, Biochemistry, and Behavior.*, *100*(4), 678-687.
[75] Sang, C. N., Ramadan, N. M., Wallihan, R. G., Chappell, A. S., Freitag, F. G. et al. (2004). LY293558, a novel AMPA/GluR5 antagonist, is efficacious and well-tolerated in acute migraine. *Cephalalgia.*, *24*(7), 596-602.
[76] Apland, J. P., Aroniadou-Anderjaska, V., Figueiredo, T. H., Green, C. E., Swezey, R. et al. (2013). Efficacy of the GluK1/AMPA receptor antagonist LY293558 against seizures and neuropathology in a soman-exposure model without pretreatment and its pharmacokinetics after intramuscular administration. *Journal of Pharmacology and Experimental Therapeutics.*, *344*(1), 133-140.
[77] Weiss, B., Alt, A., Ogden, A. M., Gates, M., Dieckman, D. K. et al., (2006). Pharmacological characterization of the competitive GLUK5 receptor antagonist decahydroisoquinoline LY466195 in vitro and in vivo. *Journal of Pharmacology and Experimental Therapeutics.*, *318*(2), 772-781.
[78] Buzzi, M. G., Bonamini, M. & Moskowitz, M. A. (1995). Neurogenic model of migraine. *Cephalalgia.*, *15*(4), 277-280.
[79] Gomez-Mancilla, B., Brand, R., Jurgens, T. P., Gobel, H., Sommer, C. et al., (2014). Randomized, multicenter trial to assess the efficacy, safety and tolerability of a single dose of a novel AMPA receptor antagonist BGG492 for the treatment of acute migraine attacks. *Cephalalgia.*, *34*(2), 103-113.
[80] Peroutka, S. J. (2014). Clinical trials update. 2013: year in review. *Headache.*, *54*(1), 189-194.
[81] Sprenger, T. & Goadsby, P. J. (2009). Migraine pathogenesis and state of pharmacological treatment options. *Bmc Medicine.*, 7.
[82] Keywood, C., Wakefield, M. & Tack, J. (2009). A proof-of-concept study evaluating the effect of ADX10059, a metabotropic glutamate receptor-5 negative allosteric modulator, on acid exposure and symptoms in gastro-oesophageal reflux disease. *Gut.*, *58*(9), 1192-1199.
[83] Marin, J. C. & Goadsby, P. J. (2010). Glutamatergic fine tuning with ADX-10059: a novel therapeutic approach for migraine? *Expert Opin Investig Drugs.*, *19*(4), 555-561.
[84] Calabresi, P., Galletti, F., Rossi, C., Sarchielli, P. & Cupini, L. M. (2007). Antiepileptic drugs in migraine: from clinical aspects to cellular mechanisms. *Trends Pharmacol Sci.*, *28*(4), 188-195.

[85] Hoffmann, J., Akerman, S. & Goadsby, P. J. (2014). Efficacy and mechanism of anticonvulsant drugs in migraine. *Expert Rev Clin Pharmacol.*, *7*(2), 191-201.

[86] Chowdhury, M., Anwar, A. U., Omar, K. H. & Majumder, S. (2013) Sodium Valproate in Migraine Prevention: Efficacy is the same as Propranolol.

[87] Silberstein, S. D. (2010). Migraine preventive treatment. *Handb Clin Neurol.*, *97*(337-354.

[88] Pelzer, N., Stam, A. H., Haan, J., Ferrari, M. D. & Terwindt, G. M. (2013). Familial and sporadic hemiplegic migraine: diagnosis and treatment. *Current Treatment Options in Neurology.*, *15*(1), 13-27.

[89] Jefferson, J. W. (2005). Lamotrigine in psychiatry: pharmacology and therapeutics. *CNS Spectr.*, *10*(3), 224-232.

[90] Lebas, A., Guyant-Marechal, L., Hannequin, D., Riant, F., Tournier-Lasserve, E. et al. (2008). Severe attacks of familial hemiplegic migraine, childhood epilepsy and ATP1A2 mutation. *Cephalalgia.*, *28*(7), 774-777.

[91] Stefani, A., Spadoni, F., Siniscalchi, A. & Bernardi, G. (1996). Lamotrigine inhibits Ca2+ currents in cortical neurons: functional implications. *European Journal of Pharmacology.*, *307*(1), 113-116.

[92] Lapidus, K. A., Soleimani, L. & Murrough, J. W. (2013). Novel glutamatergic drugs for the treatment of mood disorders. *Neuropsychiatr Dis Treat.*, *9*(1101-1112.

[93] Steiner, T. J., Findley, L. J. & Yuen, A. W. (1997). Lamotrigine versus placebo in the prophylaxis of migraine with and without aura. *Cephalalgia.*, *17*(2), 109-112.

[94] Steiner, T., Findley, L. & Yuen, A. (1997). Lamotrigine versus placebo in the prophylaxis of migraine with and without aura. *Cephalalgia.*, *17*(2), 109-112.

[95] Lampl, C., Buzath, A., Klinger, D. & Neumann, K. (1999). Lamotrigine in the prophylactic treatment of migraine aura--a pilot study. *Cephalalgia.*, *19*(1), 58-63.

[96] D'Andrea, G., Granella, F., Cadaldini, M. & Manzoni, G. C. (1999). Effectiveness of lamotrigine in the prophylaxis of migraine with aura: an open pilot study. *Cephalalgia.*, *19*(1), 64-66.

[97] Chen, W. T., Fuh, J. L., Lu, S. R. & Wang, S. J. (2001). Persistent migrainous visual phenomena might be responsive to lamotrigine. *Headache.*, *41*(8), 823-825.

[98] Pascual, J., Caminero, A. B., Mateos, V., Roig, C., Leira, R. et al. (2004). Preventing disturbing migraine aura with lamotrigine: an open study. *Headache.*, *44*(10), 1024-1028.

[99] Lampl, C., Katsarava, Z., Diener, H. C. & Limmroth, V. (2005). Lamotrigine reduces migraine aura and migraine attacks in patients with migraine with aura. *J Neurol Neurosurg Psychiatry.*, *76*(12), 1730-1732.

[100] Pelzer, N., Stam, A. H., Carpay, J. A., Vries, B. D., van den Maagdenberg, A. M. et al. (2014). Familial hemiplegic migraine treated by sodium valproate and lamotrigine. *Cephalalgia.*

[101] Kaube, H., Herzog, J., Kaufer, T., Dichgans, M. & Diener, H. C. (2000). Aura in some patients with familial hemiplegic migraine can be stopped by intranasal ketamine. *Neurology.*, *55*(1), 139-141.

[102] Afridi, S. K., Giffin, N. J., Kaube, H. & Goadsby, P. J. (2013). A randomized controlled trial of intranasal ketamine in migraine with prolonged aura. *Neurology.*, *80*(7), 642-647.

[103] Linde, M., Mulleners, W. M., Chronicle, E. P. & McCrory, D. C. (2013). Topiramate for the prophylaxis of episodic migraine in adults. *Cochrane Database Syst Rev.*, *6*(Cd010610.

[104] Silberstein, S. D., Neto, W., Schmitt, J., Jacobs, D. & Group, M-S. (2004). Topiramate in migraine prevention: results of a large controlled trial. *Arch Neurol.*, *61*(4), 490-495.

[105] Silberstein, S. D., Lipton, R. B., Dodick, D. W., Freitag, F. G., Ramadan, N. et al. (2007). Efficacy and safety of topiramate for the treatment of chronic migraine: A randomized, double-blind, placebo-controlled trial. *Headache.*, *47*(2), 170-180.

[106] Silberstein, S. D., Lipton, R. B., Dodick, D. W., Freitag, F. G., Ramadan, N. et al. (2007). Efficacy and safety of topiramate for the treatment of chronic migraine: a randomized, double-blind, placebo-controlled trial. *Headache.*, *47*(2), 170-180.

[107] Lipton, S. A. (2004). Failures and Successes of NMDA Receptor Antagonists: Molecular Basis for the Use of Open-Channel Blockers like Memantine in the Treatment of Acute and Chronic Neurologic Insults. *NeuroRx.*, *1*(1), 101-110.

[108] Xia, P., Chen, H. S., Zhang, D. & Lipton, S. A. (2010). Memantine preferentially blocks extrasynaptic over synaptic NMDA receptor currents in hippocampal autapses. *J Neurosci.*, *30*(33), 11246-11250.

[109] Parsons, C. G., Danysz, W. Quack, G. (1999). Memantine is a clinically well tolerated N-methyl-D-aspartate (NMDA) receptor antagonist--a review of preclinical data. *Neuropharmacology.*, *38*(6), 735-767.

[110] Carlton, S. M., Rees, H., Tsuruoka, M. & Willis, W. D. (1998). Memantine attenuates responses of spinothalamic tract cells to cutaneous stimulation in neuropathic monkeys. *Eur J Pain.*, *2*(3), 229-238.

[111] Medvedev, I. O., Malyshkin, A. A., Belozertseva, I. V., Sukhotina, I. A., Sevostianova, N. Y., et al. (2004). Effects of low-affinity NMDA receptor channel blockers in two rat models of chronic pain. *Neuropharmacology.*, *47*(2), 175-183.

[112] Zhang, G. H., Min, S. S., Lee, K. S., Back, S. K., Yoon, S. J. et al. (2004). Intraarticular pretreatment with ketamine and memantine could prevent arthritic pain: relevance to the decrease of spinal c-fos expression in rats. *Anesth Analg.*, *99*(1), 152-158.

[113] Johnson, J. W. & Kotermanski, S. E. (2006). Mechanism of action of memantine. *Current Opinion in Pharmacology.*, *6*(1), 61-67.

[114] Rogawski, M. A. & Wenk, G. L. (2003). The neuropharmacological basis for the use of memantine in the treatment of Alzheimer's disease. *Cns Drug Reviews.*, *9*(3), 275-308.

[115] Krymchantowski, A. & Jevoux, C. (2009). Memantine in the Preventive Treatment for Migraine and Refractory Migraine. *Headache: The Journal of Head and Face Pain.*, *49*(3), 481-482.

[116] Charles, A., Flippen, C., Romero Reyes, M. & Brennan, K. C. (2007). Memantine for prevention of migraine: a retrospective study of 60 cases. *J Headache Pain.*, *8*(4), 248-250.

[117] Spengos, K., Theleritis, C. & Paparrigopoulos, T. (2008). Memantine and NMDA antagonism for chronic migraine: a potentially novel therapeutic approach? *Headache.*, *48*(2), 284-286.

[118] Bigal, M., Rapoport, A., Sheftell, F., Tepper, D. & Tepper, S. (2008). Memantine in the preventive treatment of refractory migraine. *Headache.*, *48*(9), 1337-1342.

[119] HIS. (2013). The International Classification of Headache Disorders, 3rd edition (beta version). *Cephalalgia.*, *33*(9), 629-808.

[120] Lindelof, K. & Bendtsen, L. (2009). Memantine for prophylaxis of chronic tension-type headache--a double-blind, randomized, crossover clinical trial. *Cephalalgia.*, *29*(3), 314-321.

[121] Simpson L. L. (1981). The origin, structure, and pharmacological activity of botulinum toxin. *Pharmacol Rev.*, *33*(3), 155-188.

[122] Smuts, J. A., Schultz, D. & Barnard, A. (2004). Mechanism of action of botulinum toxin type A in migraine prevention: a pilot study. *Headache.*, *44*(8), 801-805.

[123] Berry, M. G. & Stanek, J. J. (2012). Botulinum neurotoxin A: a review. *J Plast Reconstr Aesthet Surg.*, *65*(10), 1283-1291.

[124] Binder, W. J., Blitzer, A. & Brin, M. F. (1998). Treatment of hyperfunctional lines of the face with botulinum toxin A. *Dermatologic Surgery.*, *24*(11), 1198-1205.

[125] Aurora, S. K., Dodick, D. W., Diener, H. C., DeGryse, R. E., Turkel, C. C. et al., (2014). OnabotulinumtoxinA for chronic migraine: efficacy, safety, and tolerability in patients who received all five treatment cycles in the PREEMPT clinical program. *Acta Neurologica Scandinavica.*, *129*(1), 61-70.

[126] Schwedt, T. J. (2014). Chronic migraine. *BMJ.*, *348* (g1416.

[127] Silberstein, S. D., Dodick, D. W., Aurora, S. K., Diener, H. C., DeGryse, R. E. et al. (2014). Per cent of patients with chronic migraine who responded per onabotulinumtoxinA treatment cycle: PREEMPT. *J Neurol Neurosurg Psychiatry.*

[128] Bittencourt da Silva, L., Karshenas, A., Bach, F. W., Rasmussen, S., Arendt-Nielsen, L. et al., (2014). Blockade of glutamate release by botulinum neurotoxin type A in humans: a dermal microdialysis study. *Pain Res Manag.*, *19*(3), 126-132.

[129] Cairns, B. E., Sessle, B. J. & Hu, J. W. (2001). Characteristics of glutamate-evoked temporomandibular joint afferent activity in the rat. *J Neurophysiol.*, *85*(6), 2446-2454.

[130] Cairns, B. E., Gambarota, G., Svensson, P., Arendt-Nielsen, L. & Berde, C. B. (2002). Glutamate-induced sensitization of rat masseter muscle fibers. *Neuroscience.*, *109*(2), 389-399.

[131] Cairns, B. E., Svensson, P., Wang, K., Hupfeld, S., Graven-Nielsen, T. et al. (2003). Activation of peripheral NMDA receptors contributes to human pain and rat afferent discharges evoked by injection of glutamate into the masseter muscle. *J Neurophysiol.*, *90*(4), 2098-2105.

[132] Finkel, A. G. (2011). Botulinum Toxin and the Treatment of Headache: A Clinical Review. *Headache: The Journal of Head and Face Pain.*, *51*(10), 1565-1572.

In: Modulators of Glutamatergic Signaling
Editor: Zoran M. Pavlovic
ISBN: 978-1-63483-493-3

Chapter 5

OBSESSIVE-COMPULSIVE DISORDER

***Gyula Bokor*[1,*] *and Peter D. Anderson*[2]**
[1]Staff Psychiatrist, Taunton State Hospital, Taunton, MA, US
[2]Psychopharmacology Consultant, Randolph, MA, US

ABSTRACT

Obsessive-compulsive (OCD) is a heterogeneous disorder consisting of intrusive thoughts and repetitive behavior. Serotonin is the neurotransmitter traditionally linked to OCD. The first line pharmacotherapy for OCD is serotonin reuptake inhibitors (SRIs). Often 12 weeks of treatment is needed to have an adequate response and many patients experience an inadequate response to SRIs. Research has focused on new interactions for treating OCD. Evidence of glutamatergic involvement in OCD includes genetics, radiological , and pharmacological studies. Glutamatergic drugs that have been clinically studied as a treatment for OCD include ketamine, riluzole, memantine, N-acetylcysteine, D-cycloserine, and lamotrigine.

Keywords: ketamine, riluzole, memantine, N-acetylcysteine, D-cycloserine, lamotrigine

INTRODUCTION

Obsessive-compulsive disorder (OCD) is a heterogeneous disorder presenting with the hallmarks of obsessions and compulsions [1]. The former are intrusive, recurrent, and persistent unwanted thoughts. The latter are repetitive behaviors or mental acts that an individual feels driven to perform in response to the obsession [2]. Examples of obsessions include preoccupation with germs, preoccupation with order and symmetry, fear of causing harm to others, excessive focus on religious themes, or unwarranted fear of losing items. Examples of compulsions are excessive hand washing and cleaning, and extreme rechecking

* Corresponding author: Gyula Bokor, M.D. Taunton state hospital Taunton, MA. 02780, e-mail: Gyula.bokor@icloud.com.

of locks, switches, or appliances [3]. The estimated prevalence of OCD in adults residing in the United States is 1% [4]. The estimated prevalence of OCD in children is 1 to 2% [5].

HISTORY OF OCD

Features of OCD have probably been present since humans occupied the earth. Prior to the 1700s the clergy primarily addressed OCD-related behavior [6]. Treatments of OCD have included bloodletting, laxatives and enemas [6]. In the 1700s and 1800s it was common practice to institutionalize individuals with OCD symptoms. In the 19th century physicians recognized that OCD was not a form of psychosis [7].

In the 1800s treatment included opiates and arsenic [8]. In the early 20th century the term obsessive-compulsive disorder was coined [9]. The first major breakthrough in the treatment of OCD was the development of clomipramine in the 1960s. Clomipramine is a derivative of the tricyclic antidepressant imipramine where a chloride atom is substituted for a hydrogen atom at position 3 on the ring structure, which created a potent inhibitor of serotonin reuptake and an effective treatment for OCD [10] In contrast; imipramine possesses minimal activity on serotonin reuptake and is ineffective treatment of OCD. Exposure and response prevention therapy (ERP) for OCD was developed in the 1960s and 1970s [11]. ERP is the preferred nonpharmacological treatment for OCD. ERP works in large part because it causes fear extinction. The Diagnostic and Statistical Manual, 4th edition classified OCD as an anxiety disorder [12]. OCD was placed in a chapter on obsessive-compulsive and related disorders; hence OCD is no longer classified as an anxiety disorder.

RELATED CONDITIONS

Conditions related to OCD include hoarding disorder, trichotillomania (hair-pulling disorder), excoriation (skin picking) disorder, body dysmorphic disorder, and substance/medication induced obsessive-compulsive and related disorder, and obsessive-compulsive and related disorder due to another medical condition [2]. Hoarding disorders manifests as persistent difficulty with discarding possessions regardless of value. Trichotillomania is the recurrent pulling out of one's hair, producing hair loss. Excoriation is recurrent skin picking resulting in skin lesions. A preoccupation with one or more perceived flaws or defects in physical appearance that is not noticeable or minimal to others is termed body dysmorphic disorder. A number of drugs can produce obsessive-compulsive like behavior including amphetamines, cocaine, methylphenidate, second generation anti-psychotics, and dopaminergic agonists (especially agonists of the D3 receptor) [13, 14, 15, 16]. Neurological disorders including Huntington's Disease and strokes can produce obsessive-compulsive symptoms [17].

Obsessive-compulsive personality disorder (OCPD) consists of a pervasive pattern of preoccupation with orderliness, perfection, and mental and interpersonal control to the degree that openness, flexibility, and efficiency are impaired [18]. In OCPD, the individual is obsessed with details, rules, order, lists, organization, methodology, or schedules to the degree that the major point of the activity is lost. The manifestations of OCPD are egosyntonic, the affected individual finds the behavior correct and suitable. Whereas, the manifestations of OCD are egodystonic, the affected individual finds the behavior inappropriate and stressful. Persons with OCPD typically obtain pleasure with organizing and controlling. In certain professions such as pharmacy (especially for dispensing pharmacists), surgeons, and accountants a degree of obsessive-compulsiveness is necessary to provide optimal occupational function.

Comorbidity with OCD is common. OCD symptoms have been reported in 8% to 26% of patients with schizophrenia [17]. This type of clinical presentation may represent a distinct type of schizophrenia called "obsessive compulsive schizophrenia" rather than a true comorbidity [19]. Tics and Tourette's Syndrome are a frequent comorbidity with OCD [20]. Complex motor tics can be confused with compulsions [3]. A meta-analysis found rates of OCD in 17.4% of individuals with autism spectrum disorder (ASD) [21]. The repetitive behavior commonly manifested in ASD has been thought to be related to OCD. However, the lack of improvement in repetitive behaviors with SSRIs suggests that that repetitive behavior from ASD is unrelated to OCD [22]. The repetitive behaviors in ASD may be general relaxing activity rather than a response to an obsession [22].

NEUROBIOLOGY OF OCD

The neurobiology of OCD is very complex with numerous anatomical areas and functional pathways involved in the pathophysiology. The basal ganglia is the most implicated brain structure with OCD [23]. Other parts of the brain associated with OCD include the orbitofrontal cortex, anterior cingulated gyrus, parietal lobe, superior temporal gyrus, and hippocampus [24, 25]. OCD may be due to malfunctions in the cortico-striato-thalamo-cortical circuit in the brain [26, 27]. Another hypothesis is that the cortico-striatal-cerebellar circuits are disrupted in OCD [28]. Glutamate is active in these circuits and most areas of the brain.

Neurotransmitters implicated in OCD include serotonin, dopamine, and glutamate [29, 30, 31]. Serotonin has the most evidence in the pathophysiology of OCD [29]. The serotonin hypothesis is based largely on the positive response to SSRIs and clomipramine [29]. Case reports of improvement or aggravation of OCD with second generation antipsychotics provide evidence that dopamine and/or serotonin contributes to OCD [14, 15]. Dopamine agonists and reuptake inhibitors have been reported to exacerbate or cause OCD symptoms [13, 16]. The lack of improvement in OCD with tricyclic antidepressants suggests no major role for norepinephrine in OCD.

Research is providing more insight on the role of glutamate in the pathophysiology of OCD [31]. Neuroimaging research, animal studies, clinical trials, case reports, and pharmacological studies provide evidence of the role of glutamate in OCD [3]. Reduced serotonergic inhibition in the striatal areas, orbito-frontal cortex, and thalamus may result in excess glutamatergic activity [29]. Reduced glutamatergic activity in the anterior cingulate cortex (ACC) has been reported in children with OCD [32]. The reduced glutamatergic activity in the ACC results in phasic overactivity in the striatum and orbitofrontal cortex [32]. Two small studies found higher glutamate levels in the cerebrospinal fluid of patients with OCD than healthy controls [33, 34]. The first study consisted of 21 drug-naïve subjects with

OCD and 18 healthy controls [33]. The second study had 23 drug-naïve subjects with OCD and 23 controls [34]. Neither study provides direct insight into the mechanism of glutamate in OCD.

Genetics

Many genes have been linked to OCD including genes impacting glutamate. SLC1A1 codes for the EAAC1 protein, a transporter of glutamate. SLC1A1 has been linked to childhood-onset and adult-onset OCD [35, 36]. Shmelkov et al., found that that knock out of the SLITK5 gene, a regulator of a neuron specific transmembrane protein, produced excessive grooming behaviors (one laboratory model of compulsive behavior) and anxiety in mice [37]. Specific neurobiological findings included overactivity of the orbitofrontal area, reduced volume in the striatum, and decreased dendrite complexity. Down regulation of glutamate receptors was found in the striatum of the laboratory mice [37]. No studies have linked the SLITK5 gene to OCD in humans.

Deletion of the SAPAP3 gene produced excessive grooming behavior and anxiety in laboratory mice [38]. SAPAP3 is involved with glutamate-mediated communication in the cortex-striatum and is needed for the protein structure of the scaffolding to maintain the integrity of the synapse, including the receptors, especially in excitable post synaptic neurons. A case-control study examined the association of SAPAP3 in 44subjects with OCD and trichotillomania, 44 cases of OCD without trichotillomania (TTM), and 178 controls. SAPAP3 variants were found in 4.2% of the TTM/OCD patients but only in 1.1% of controls [39]. Boardmen et al., found an association with early-onset OCD and SAPAP3 variants but not with trichotillomania or OCD [40]. The group consisted of 172 subjects with OCD, 45 with trichotillomania, and 153 controls. The OCD collaborative genetics study found SAPAP3 associated with pathological grooming in humans but not OCD [41]. SAPAP3 normally inhibits endocytosis of AMPAR receptors. An increase in endocytosis results in decreased AMPAR receptor activity [42]. Blockade of the melanocyte 4 receptor (MC4R), either genetically or pharmacologically, rectifies compulsive behavior from SAPAP3 deletion in mice [43]. MC4R receptor antagonists have not been studied in humans with OCD. No MC4R receptors antagonists are commercially available in the United States.

The GRIN2B gene has been linked to some cases of OCD [44, 45]. This gene regulates N2B subunit of NMDA receptors which is the agonist binding site for glutamate at the primary excitatory receptor in the mammalian brain. Two genetic studies in humans suggest the GRIK2, a gene for a kainite receptor, is related to OCD [46, 47]. Kainate receptors are also ionotropic receptors of glutamate but are less understood than NMDA or AMPA receptors.

Neuroimaging Studies

Decreased thalamic glutamate levels were found in unmedicated adults with OCD by proton magnetic resonance spectroscopy (H MRS) [48]. A study of proton magnetic resonance spectroscopy found no significant differences in striatal subregions in adults with adults compared to controls [49]. A functional MRI test did not find evidence of glutamate malfunction in the anterior cingulated [50]. A pediatric study using H MRS found no difference in the glutamine in patients with OCD compared to healthy controls [51]. However, the researchers found different glutamatergic activity based on the severity of illness [51]. H MRS found no glutamate abnormalities in the prefrontal cortex of adults [52]. Glutamate levels were positively correlated with symptom severity based on H MRS and the Yale Brown obsessive-compulsive scale in a study of adults with OCD [53]. Symptom severity was correlated by glutamate levels in the anterior cingulated as indicated by H MRS [54]. An earlier study using H MRS found an association with GRIN2B polymorphisms and reduced anterior cingulated concentration in pediatric patients with OCD [44]. A small clinical trial evaluated the effects of paroxetine on glutamate levels in the caudate [55]. The study included 11 pediatric patients from 8 to 17 years of age with OCD. Glutamate levels were measured at baseline and after 12 weeks of monotherapy with paroxetine. The glutamate levels decreased after 12 weeks of treatment. The decreased was associated with improvement of symptoms. A case report states that glutamate levels in the caudate remained decreased even after the paroxetine was stopped [56].

What explains the range of differences in the neuroimaging studies? Perhaps the low number of studies presents a lack of statistical power to detect differences. The role of glutamate may be best understood in terms of interactions with other neurotransmitters than by itself. Glutamate levels may be more relevant in certain genetic populations. Further research may answer these questions.

PHARMACOTHERAPY

The selective serotonin reuptake inhibitors (SSRIs) are the primary pharmacotherapy of OCD [57]. Other common pharmacotherapies are clomipramine and serotonin norepinephrine reuptake inhibitors [3]. All three of types of drugs are serotonin reuptake inhibitors (SRI). An estimated 40 to 60% of patients have an inadequate response to this first line pharmacological treatment [58]. Definitions of refractory OCD vary from study to study. No well-established effective treatments for are available refractory OCD. Treatment of refractory OCD includes augmentation with antipsychotics, pindolol, mirtazapine, lithium, and buspirone [3]. In addition, 12 week or more is often required to achieve an adequate response with SRIs which is longer than the 6 to 8 weeks needed for improvement in depression. More treatment options are needed and this includes drugs with glutamatergic activity. Ketamine, riluzole, memantine, N-acetylcysteine, D-cycloserine, and lamotrigine are glutamatergic drugs that have been explored as treatments for OCD [31].

Ketamine is a dissociative anesthetic with the primary mechanism of action of a NMDA antagonist but also has activity on dopamine, norepinephrine and serotonin [59]. The effects of ketamine in OCD were first reported in a case report of a 24 year old woman with treatment refractory OCD [60]. Ketamine was administered at a dose of 0.5mg/kg intravenously over 40 minutes. She reported entire resolution of obsessions before completion of the infusion. The obsessions partially remerged a few hours of the infusion. The obsessions did not return to baseline until 7 days after the infusion [60].

An open-label study of ketamine was conducted in 10 subjects with treatment-refractory OCD [61]. Eight participants were on serotonergic drugs, two also on atypical antipsychotics, and two were unmedicated. Ketamine was administered at dose of 0.5mg/kg intravenously over 40 minutes. This is much lower than the 2mg/kg used in emergency procedures [62]. Clinical ratings used include the Clinical Global Impression scale and the Yale Brown Obsessive-Compulsive Scale (Y-BOCS), which is the most common rating scale used in studies of OCD. The Y-BOCS was modified for this study to account for the frequent time intervals. The rating scales were assessed at baseline; 1, 2, and 3 hours; and 1,2,3,5 and 7 days post infusion. A treatment response was defined as an improvement in 35% or more in the Y-BOCS. The Clinician Administered Dissociative States Scale was conducted to determine the degree of dissociation. An acute improvement was seen 1 to 3 hours after ketamine. None of the subjects had a treatment response during days 1 to 7 days but showed mild improvement in the symptoms.

A place-controlled randomized crossover trial of ketamine was conducted in 15 unmedicated adults with OCD [63]. Participants were required to have "near-constant obsessions." Subjects received two intravenous infusions, one of ketamine (0.5mg/kg) and one of saline, administered a minimum of a week apart. The OCD visual analog scale was used to detect acute changes in OCD symptoms. The Y-BOCS was assessed and a treatment response was defined as 35% decrease in OCD symptoms. Eight patients met the criteria for treatment response. An inherent weakness in a placebo study involving ketamine is that its psychogenic effects are so robust that it is difficult to have a true blind.

Several factors could describe the differences in the two studies. In the first study most of the patients were on psychoactive active medications. The serotonergic effects may have mitigated the glutamate changes from the ketamine. The second controlled study required "nearly constant" intrusive obsessions and this may represent a subtype of OCD that responds to ketamine. The first study had a mid-infusion dissociation level of 1.4 on the CADSS and the second study had a 22. Dissociation may be needed to have an antiobsessional response to ketamine or dissociation may be a surrogate marker of a general response to ketamine. The first study had refractory patients so they have had a track record of poor response to medications. Both studies were small so the differences in efficacy could be chance alone.

How is ketamine helpful in OCD? Ketamine acutely increases glutamate in the prefrontal cortex disrupting dopamine neurotransmission [64]. This mechanism may explain the acute relieve of OCD symptoms but not the seven days of improvement. Ketamine is known to increase BDNF which stimulates neural growth and plasticity [65]. Laboratory studies indicate that ketamine activates the mammalian target of rapamycin, which is needed for cellular proliferation [65].

Riluzole is indicated for amyotropic lateral sclerosis and the mechanism of action includes sodium channel blockade, reduction of glutamate release, prevention of glutamate receptor hypofunction, and increasing glutamate uptake [66]. The effectiveness of riluzole was first reported in 2005 from an open-label trial consisting of 13 subjects with refractory OCD [67]. The subjects had inadequate response to SRI which was defined as a Y-BOCS score above 16 after 8 weeks of pharmacotherapy. The main study outcome was improvement on the Y-BOCS. Riluzole was started at 50mg bid. All of the subjects had concomitant SRIs and 4 were on atypical antipsychotics. The mean Y-BOCS was 30.7 and decreased to 17.7 by the end of the study (12 weeks). Weaknesses of the study include an open-label and no placebo control. The mean Y-BOCS 17.7 at the end was still higher than the above 16 that defined inadequate response to SRI. The study was initially intended to be 6 weeks but it was later extended to 9 and then to 12 weeks to allow further improvement. Eight weeks of SRI therapy may have been an inadequate duration to have maximal improvement.

An open label study evaluated riluzole in six children with refractory OCD [68]. Two subjects had no improvement after 12 weeks. The other four subjects had a 46% decrease in the children Y-BOCS.

Only one placebo-controlled study of riluzole for OCD was found with a Medline search. The study included 60 children or adolescents with OCD including 17 with comorbid ASD [69]. Subjects needed to be on concomitant medications for 2 months. Treatment response was defined as a 30% reduction in the Y-BOCS scores. Subjects received 50mg of riluzole twice a day or placebo for 12 weeks. The mean change in Y-BOCS score was 21% for riluzole and 19% for placebo. The percentage of responders was identical with riluzole and placebo, 7%. Subset analysis in the patients with ASD found no differences. No reduction in repetitive behavior in the subjects with ASD was shown with riluzole. This is consistent with the notion that repetitive behaviors in ASD are etiologically distinct from compulsions as described above.

Memantine is an antagonist of glutamate at NMDA receptors indicated for the treatment of Alzheimer's disease [70]. Memantine also is an antagonist at the 5-HT3 receptor which may contribute to antidepressant properties [71]. Memantine is also has agonist actions at the D2 receptor [72]. A single-blinded case-control study evaluated memantine for the treatment of severe OCD [73]. The study included 44 subjects receiving pharmacological treatment and CBT at an intensive residential treatment program for OCD. Half the subjects received memantine as an augmenting agent. The primary outcome measure was the Y-BOCS. A responder was considered a 25% improved and a marked responder had a 50% or greater reduction. The mean decrease in Y-BOCS was 7.2 for the cases and 4.6 for the controls. Overall the control group had a 27% decrease and the control group had a 16.5% decrease. The starting dose was 5mg and the mean final dose was 18.0 mg. Two subjects were responders and 6 subjects were marked responders in the memantine group.

A randomized double-blind placebo-controlled study also found that memantine improved OCD symptoms [74]. The study included 42 patients with non-refractory OCD. All patients received fluvoxamine 100mg daily for the first four weeks followed by 200mg for the rest of the study. The subjects randomly assigned to receive memantine 20mg daily or placebo. Y-BOCS was measured at baseline, 2, 4, 6, and 8 weeks. The main outcome was the difference in score change from the two groups. Secondary outcome measures were partial response rates (> 25% reduction in Y-BOCS), complete response (> 35% reduction in Y-BOCS), and remission (score < 16). All of the patients in the memantine group had either partial or complete response. Only six patients (32%) in placebo group had partial or complete response. Seventeen patients (89%) in the memantine group six patients (32%) in the placebo group obtained remission. Weaknesses in the study are that the patients were not refractory to treatment. A Y-BOCS score of 16 still represents clinically significant symptoms of OCD and thus not the best criteria for remission. The key finding is that memantine improves the response time to improvement.

Another placebo-controlled double-blind study evaluated adjuvant memantine in patients with refractory OCD [75]. The study included 40 patients with ODC who received an SSRI or clomipramine at therapeutic doses for a minimum of 12 weeks. A minimum score of 21 on the Y-BOCS was required. The Y-BOCS was obtained baseline, and at weeks 4, 8 and 12. A full response defined as a 35% improvement in Y-BOCS, a partial response a 25% to 35% improvement, and no response 25% or less of improvement. The memantine group had a mean reduction of 9.29 compared to 4.46 with placebo at 12 weeks. Nine patients (60%) in the memantine group and none in placebo group obtained a full response. One subject (6.6%) in the memantine group and 11 (78.%) in the placebo group were nonresponders. None of the memantine studies evaluated in vivo glutamatergic activity in the brain.

N-acetylcysteine (NAC) supplies cysteine as a precursor to glutathione, the primary endogenous antioxidant in the body. The most common uses of NAC are the treatment of acetaminophen poisoning and as a mucolytic. NAC is becoming used more often in psychiatry [76]. With oral administration, NAC undergoes deacetylation in the liver producing cysteine, an amino acid. Further down the biochemical pathway cysteine is used in the synthesis of glutathione or converted to cystine. Research is conflicting if cysteine is able cross the blood brain barrier [77]. NAC activates the cystine/glutamate exchange, found on glial cells, causing the reverse transport of glutamate into the extrasynaptic space. This results in stimulation of inhibitory metabotropic glutamate receptors (mGluR2/3) and reducing synaptic release of glutamate [78]. Oxidative stress has not been extensively studied in OCD. However, at this point the antioxidant properties of NAC cannot be ruled out as a contributing factor to the benefits in OCD [79].

The effectiveness of NAC for OCD was first documented in a case report by Lafeleur et al. in 2005 [80]. NAC improved marble-burying behavior in mice, one laboratory model for OCD [81]. A placebo-controlled double-bind study evaluated NAC as adjuvant therapy for OCD [82]. The study included 48 subjects with OCD that failed to respond to a minimum of 12 weeks of an SSRI or clomipramine. Subjects in the active treatment arm received an initial dose of 600mg daily of NAC. Doses were doubled weekly to a maximum dose of 2400mg per day. The Y-BOCS was conducted at baseline and every four weeks. The primary outcome measure was the difference in baseline to endpoint and the rate of response in each group at the end of the trial. An improvement of 35% or greater was defined as full response, a 25% to 35% decrease was a partial response, and less than 25% was considered no response. Ten patients (52.6%) in the NAC group experienced full response. The actual numbers of full responders on placebo was not reported but from the graph appears to be approximately 10%. The main side effect was gastrointestinal upset.

Grant et al. evaluated NAC for trichotillomania in a double-blind, placebo-controlled study [83]. Fifty subjects with trichotillomania were recruited. Psychotropic medications were allowed if the dose was stable for the previous three months. The primary measures of outcome were Massachusetts General Hospital Hair Pulling Scale (MGH-HPS) and Psychiatric Institute Trichotillomania Scale (PITS) which were assessed at baseline and every three weeks. NAC was started at 1200mg per day and after six weeks the dose was increased to 2400mg per day. The placebo group had a baseline MGH-HPS of 16.7 and decreased to 16.00 at the end of the study. In contrast the NAC group had a 17.6 baseline and 10.4 at the end. The placebo group had a baseline PITS of 23 and an end score of 18.7. The NAC group had a baseline PITTS of 21.2 and an end score of 11.8. Eleven patients (44%) in the NAC arm had a 50% or greater reduction in the MGH-HPS compared to 0% in the placebo arm. Quality of life was no better in the NAC group compared to placebo. Eleven patients did not respond to NAC. No significant side effects were reported.

Cycloserine is an antibiotic that serves as a partial agonist at the NMDA receptor in the basolateral amygdala. Chasson et al. studied the D-cycloserine as adjuvant treatment with exposure and response prevention therapy (ERP) for OCD [84]. The study consisted of 22 adults with OCD that received ten sessions of ERP. Half the subjects received placebo and half received 100mg of D-cycloserine one hour before the ERP. The primary measure of efficacy was Y-BOCS. D-cycloserine appeared to speed the response of therapy by 2.3 times based on ratio of the slopes.

A preliminary study in 30 pediatric patients found no significant benefits from the addition of D-cycloserine (0.7mg/kg) to cognitive behavior therapy (which included ERP) [85]. A meta-analysis found no overall benefit of d-cycloserine as an augmentation for behavioral therapy [86]. However, benefits may be seen midtreatment. In addition, subjects that received d-cycloserine before the ERP rather than after ERP had the best improvement.

A large placebo-controlled study evaluated the efficacy of d-cycloserine as add-on therapy to ERP [87]. The study included 128 adults with OCD. Concurrent psychotropic medications were permitted if at stable doses for the preceding two months. Each subject received ERP for 12 weeks. Patients were randomized to receive either placebo or 50mg d-cycloserine before each ERP session. The primary outcome measure was Y-BOCS. Responders were defined by a 35% reduction in Y-BOCS and remission was defined as a 12 or lower score. The d-cycloserine arm had a response rate of 63% and a remission rate of 48% at follow-up. The placebo arm had a response rate of 63% and a remission rate of 50% at follow-up. In the d-cycloserine arm, patients on antidepressants had worse outcomes based on Y-BOCS than patients without antidepressants. No side serious side effects were reported from D-cycloserine. The lack of improvement in patients on antidepressants may be explained by the down regulation of NMDA receptors by antidepressants [88]. As discussed above, ketamine had minimal effects on OCD symptoms in patients being treated with serotonergic drugs [61].

Lamotrigine is an anticonvulsant that inhibits the release of glutamate. A small placebo-controlled study of 33 subjects found lamotrigine 100mg effective as an augmentation to SRIs [89]. Patients received at least 12 weeks of an SRI. A full response was defined as 35% reduction in the Y-BOCS. Seven of the 17 patients receiving lamotrigine had a full response compared to none in the placebo group. The most common side effects were drowsiness and headache. One patient had a mild rash. A chart review found lamotrigine effective in 22 adult patients resistant to SRI [90]. Treatment resistance was defined as partial response or no response to 14 weeks of either an SSRI or clomipramine. The average number of failed SRI trials was 2.37. The primary outcome measure was the Y-BOCS. The mean baseline Y-BOCS score was 28.87. The mean dose of lamotrigine was 150mg. The mean decline in Y-BOCS was 67.23%. The mean time for a 25% reduction in Y-BOCS was 9.2 weeks. Side effects reported include headache, sedation, and fatigue. One patient had a minor skin rash that was determined not be Stevens-Johnson Syndrome. The limitations of this study include the retrospective review and the small sample size.

Areas of Future Research and Conclusion

Evidence of glutamate's role includes genetics, radiological procedures and pharmacological trials. The most dramatic response is with ketamine. Memantine and D-cycloserine may shorten the response time. Mechanisms that drugs may improve OCD symptoms include

NMDA blockage, preventing glutamate receptor hypofunctioning, and inhibition of glutamate release. None of the clinical trials correlated an improvement in OCD symptoms with glutamatergic activity in the brain (e.g., neuroimaging studies) so improvement may be related to other pharmacological mechanisms. Most of the published studies contained low numbers of subjects. Within studies some patients have dramatic improvements whereas others have no significant improvement. This suggests that pharmacogenetic factors are likely involved. Another area of potential research is determining the specific role of glutamatergic drugs. Should glutamatergic modulators be used for augmentation or primary therapy? Antidepressants or antipsychotics may mitigate or negate the benefits of certain glutamatergic therapies. If glutamatergic are used for patients refractory to SRIs, should it be used before or after a second generation antipsychotic is tried. A head-to-head trial of a glutamatergic drug to a second generation antipsychotic for patients with refractory OCD may provide interesting data. OCD is a heterogeneous disorder and perhaps certain subtypes respond best to glutamatergic interventions. Further research is needed to answer these questions.

REFERENCES

[1] International OCD Foundation. About OCD – What is OCD. http://iocdf.org/about-ocd/. Accessed April 30, 2015.

[2] American Psychiatric Association. Obsessive-Compulsive and Related Disorders. Diagnostic and Statistical Manual of Mental Disorders, 5th edition. Washington, DC: *American Psychiatric Association*, 2013, 235 – 264.

[3] Bokor, G. & Anderson, P. D. (2014). Obsessive-compulsive disorder. *J Pharm.Pract.*, *27*, 116 – 30.

[4] National Institute of Mental Health. Obsessive-Compulsive Disorder Among Adults. www.nimh.nih.gov/ health. Accessed April 30, 2015.

[5] American Academy of Child and Adolescent Psychiatry. Practice Guidelines for http://www.jaacap.com/article/S0890-8567 (11)00882-3/pdf. Accessed April 30, 2015.

[6] Anonymous. Physicians Take Over: The 1700s or so. http://www.ocdhistory.net/nutshell/earlymed.html. Accessed April 30, 2015.

[7] Anonymous. The Growth of Asylums. http:// www.ocdhistory.net/nutshell/asylums.html. Accessed April 30, 2015.

[8] Anonymous. Medical Classification Debates the 1800s. http://www.ocdhistory.net/nutshell/1800s.html. Accessed April 30, 2015.

[9] Anonymous. The Freudian Revolution the early 20th Century. http://www.ocdhistory.net/nutshell/freudetc.html. Accessed April 30, 2015.

[10] The database and ontology of chemical entities of biological interest. CHEBI: 47780. http:// www.ebi.ac.uk/ chebi/ searchId.do? chebiId=CHEBI:47780. Accessed April 30, 2015.

[11] Anonymous. Developments Later in the 20th Century. http://www. ocdhistory.net/nutshell/recent.html. Accessed April 30, 2015.

[12] American Psychiatric Association. Anxiety Disorders. *Diagnostic and Statistical Manual of Mental Disorders, 4th edition.* Washington, DC: *American Psychiatric Association*, 1994, 393 – 444.

[13] Kouris, S. (1998). Methylphenidate-induced obsessive-compulsiveness. *J Am Acad Child Adolesc Psychiatry.*, *37*, 135.

[14] Lemke, N. T. & Bustillo, J. R. (2013). Clozapine-induced obsessive-compulsive symptoms in bipolar disorder. *Am J Psychiatry.*, *170*, 930.

[15] Desarkar, P., Das, A. & Nizamie, S. H. (2007). Aripiprazole-induced obsessive-compulsive disorder: a report of 2 cases. *J Clin Psychopharmacol.*, *27*, 305 – 6.

[16] Kelley, B. J., Duker, A. P. & Chiu, P. (2012). Dopamine agonists and pathologic behaviors. *Parkinson Dis.*, 2012, 2012:063631. Doi:10:1155/ 2012/603631.

[17] Pallanti, S., Grassi, G. & Sarrecchia, E. D. (2011). Obsessive-compulsive disorder comorbidity: clinical assessment and therapeutic implications. *Front Psychiatry.*, *70*, 1 – 11.

[18] American Psychiatric Association (2013). *Personality Disorders, Diagnostic and Statistical Manual of Mental Disorders, 5th edition.* Washington, DC: American Psychiatric Association, 645 – 684.

[19] Berman, I. (1999). Is there a subtype of obsessive-compulsive schizophrenia? *Psychiatric Annals.*, *29*, 506.

[20] Jankovic, J. (2001). Tourette's Syndrome. *N Eng J Med.*, *345*, 1184 – 92.

[21] van Steensel, F. J., Bogels, S. M. & Perrin, S. (2011). Anxiety disorders in children and adolescents with autistic spectrum disorders: a meta-analysis. *Clin Child Fam Psychol Rev.*, *14*, 302 – 17.

[22] Anderson P. D. & Bokor, G. (2015). Autism spectrum disorders. *J Pharm Pract.* (accepted for publication).

[23] Ring, H. A. & Serra-Mestres, J. (2002). Neuropsychiatry of the basal ganglia. *J Neurol Neurosurg Psychiatry.*, *72*, 12 -21.

[24] Graybiel, A. M. & Rauch, S. L. (2000). Toward a neurobiology of obsessive-compulsive disorder. *Neuron.*, *28*, 343-7.

[25] Maia, T. V., Cooney, R. E. & Peterson, B. S. (2008). The neural bases of obsessive-compulsive disorder in children and adults. *Dev Psychopathol.*, *20*, 1251- 83.

[26] Marsh. R., Maia, T. V. & Peterson, B. S. (2009). Functional disturbances within frontostriatal circuits across multiple childhood psychopathologies. *Am J Psychiatry.*, *166* , 664 – 74.

[27] Ahmari, S. E., Spellman, T., Douglass, N. L., Kheirbek, M. A., Simpson, H. B., Deisseroth, K. et al. (2013). Repeated Coritoc-striatal stimulation generates persistent OCD-like behavior. *Science.*, *340*, 1234-9.

[28] Anticevic, A., Hu, S., Zhang, S., Savic, A., Billingslea, E., Wasylink, S. et al. (2014). Global resting-state functional magnetic resonance imaging analysis identifies frontal cortex, striatal, and cerebellar dysconnectivity in obsessive-compulsive disorder. *Biol Psychiatry.*, *75*, 595 – 605.

[29] Goddard, A. W., Shekhar, A., Whiteman, A. F. & McDougle, C. J. (2008). Serotonergic mechanisms in the treatment of obsessive-compulsive disorder. *Drug Discov Today.*, *13*, 325 – 32.

[30] Hesse, S., Muller, U., Lincke, T., Barthel, H., Villmann, T., Angermeyer, M. C. et al. (2005). Serotonin and dopamine transporter imaging in patients with obsessive-compulsive disorder. *Psychiatry Res.*, *140*, 63 – 72.

[31] Pittenger, C., Bloch, M. H. & Williams, K. (2011). Glutamate abnormalities in obsessive compulsive disorder: neurobiology, pathophysiology, and treatment. *Pharmacol Ther.*, *132*, 314 – 32.

[32] Rosenberg, D. R., Mirza, Y., Russell, A., Tang, J., Smith, J. M., Banerjee, S. P. et al. (2004). Reduced anterior cingulate glutamatergic concentrations in childhood OCD and major depression versus healthy controls. *J Am Acad Child Adolesc Psychiatry.*, *43*, 1146 – 53.

[33] Chakrabarty, R., Bhattacharyya, S., Christopher, R. & Khanna, S. (2005). Glutamatergic dysfunction in OCD. *Neuropsychopharmacology.*, 2005, 30, 1735 – 40.

[34] Bhattacharyya, S., Khanna, S., Chakrabarty, K., Mahadevan, A., Christopher, R. & Shankar, S. K. (2009). Antibrain autoantibodies and altered excitatory neurotransmitters in obsessive-compulsive disorder. *Neuropsychopharmacology.*, *34*, 2489 – 96.

[35] Veenstra-VanderWeele, J., Kim, S. J., Gonen, D., Hanna, G. L., Leventhal, B. L. & Cook, E. H. (2001). Genonomic organization of the SLC1A1/EAAC1 gene and mutation screening in early onset obsessive-compulsive disorder. *Mol Psychiatry.*, *6* , 160 – 167.

[36] Wendlund, J. R., Moya, P. R., Timpano, K. R., Anavitarte, A. P., Kruse, M. R. & Wheaton, M. G. (2009). A halotype containing quantitative trait loci for SLC1A1 gene expression and its association with obsessive-compulsive disorder. *Arch Gen Psychiatry.*, *66*, 408 – 16.

[37] Shmelkov, S., Hormigo, A., Jing, D., Proenca, C. C., Bath, K. G., Midle, T. et al. (2010). Slitrk5 deficiency impairs corticostriatal circuitry and leads to obsessive-compulsive-like behavior in mice. *Nat Med.*, *16*, 598 – 2.

[38] Welch, J. M., Lu, J., Rodriguiz, R. M., Trotta, N. C., Peca, J., Ding, J. D. et al. (2007). Cortico-striatal synaptic defects and OCD-like behaviors in SAPAP3-mutant mice. *Nature.*, *448*, 894 – 900.

[39] Zuchner, S., Wendland, J. R., Ashley-Koch, A. E., Collins, A. L., Tran-Viet, K. N., Quinn, K. et al. (2009). Multiple rare SAPAP3 missense variants in trichotillomania and OCD. *Mol Psychiatry.*, *14*, 6 – 9.

[40] Boardmen, L., van der Merwe, L., Lochner, C., Kinnear, C. J., Seedat, S., Stein, D. J. et al. (2001). Investigating SAPAP3 variants in the etiology of obsessive-compulsive disorder and trichotillomania in the South African white population. *Compr Psychiatry.*, *52*, 181 – 7.

[41] Bienvenu O. J., Wang, Y., Shugart, Y. Y., Welch, J. M., Grados, M. A., Fyer A. J. et al. (2009). SAPAP3 and pathological grooming in humans: results from the OCD collaborative genetics group. *Am J Med Genet B Neurosychiatr Genet.*, *150B*, 710 – 20.

[42] Wan, Y., Feng, G. & Calakos, N. (2011). Sapap3 deletion causes mGluR5-dependent silencing of AMPAR synapses. *J Neurosci.*, *31*, 16685-91.

[43] Xu, P., Grueter, B. A., Britt, J. K., McDaniel, L., Huntington, P. J., Hodge, R. et al. (2013). Double deletion of melanocortin 4 receptors and SAPAP3 corrects compulsive behavior and obesity in mice. *Prot Natl Acad Sci USA.*, *110*, 10759-64.

[44] Arnold, P. D., Macmaster, F. P., Richter, M. A., Hanna, G. L., Sicard, T. & Burroughs, E. (2009). Glutamate receptor gene (GRIN2B) associated with reduced anterior cingulated glutamatergic concentration in pediatric obsessive-compulsive disorder. *Psychiatry Res.*, *172*, 136 – 9.

[45] Alonso, P., Gratacos, M., Segalas, C., Escaramis, G., Real, E., Bayes, M. et al. (2012). Association between the NMDA glutamate receptor GRINB2B gene and obsessive-compulsive disorder. *J Psychaitry Neurosci.*, *37*, 273 – 281.

[46] Delorme, R., Krebs, M. O., Chabane, N., Roy, I, Millet, B., Mouren-Simeoni, M. C. et al. (2004). Frequency and transmission of glutamate receptors GRIK2 and GRIK3 polymorphisms in patients with obsessive compulsive disorder. *Neuroreport.*, *15*, 699 – 702.

[47] Sampaio, A. S., Fagerness, J., Crane, J., Delorme, R., Pauls, D. L., Stewart, S. E. et al. (2011). Association between polymorphisms in GRIK2 gene and obsessive-compulsive disorder: a family-based study. *CNS Neurosci Ther.*, *17*, 141 -7.

[48] Zhu, Y., Fan, Q., Han, X., Zhang, H., Chen, J., Wang, Z. et al. (2015). Decreased thalamic glutamate level in unmedicated adult obsessive-compulsive disorder patients detected by proton magnetic resonance spectroscopy. *J Affect Disord.*, *178*, 193 – 200.

[49] Simpson, H. B., Kegeles, L. S., Hunter, L., Mao, X., Van Meter, P., Xu, X. et al. (2015). Assessment of glutamate in striatal subregions in obsessive-compulsive disorder with proton magnetic resonance spectroscopy. *Psychiatry Res.*, *232*, 65 – 70.

[50] Brennan, B. P., Tkachenko, O., Schwab, Z. J., Juelich, R. J., Ryan, E. M. et al. (2015). An examination of rostral anterior cingulated cortex function and neurochemistry in obsessive-compulsive disorder. *Neuropsychopharmacology.*, 2015, 9. Doi 10.1038/npp.2015.36.

[51] Ortiz, A. E., Ortiz, A. G., Falcon, C., Morer, A., Plana, M. T., Bargallo, N. & Lazaro, L. (2015) 1H-MRS of the anterior cingulated cortex in childhood and adolescent obsessive-compulsive disorder: a case-control study. *Eur Neuropsychopharmacol.*, *25*, 60 – 8.

[52] Simpson, H. B., Shungu, D. C., Bender, J., Mao, X., Xu, X., Shlifstein, M. & Kegeles, L. S. (2012). Investigation of cortical glutamate and γ-aminobutryic acid in obsessive-compulsive disorder by proton magnetic resonance spectroscopy. *Neuropsychopharmacology.*, *37*, 2684 – 92.

[53] Starck, G., Ljungberg, M., Nilsson, M., Jonsson, L., Lundberg, S., Ivarsson, T. et al. (2008). A 1H magnetic resonance spectroscopy study in adults with obsessive compulsive disorder: relationship between metabolite concentrations and symptom severity. *J Neural Trans.*, *115*, 1051 – 62.

[54] Yucel, M., Wood, S. J., Wellard, R. M., Harrison, B. J., Fornito, A. et al. (2008). Anterior cingulated glutamate-glutamine levels predict symptom severity in women with obsessive-compulsive disorder. *Aust N Z Psychiatry.*, *42*, 467-77.

[55] Rosenberg, D. R., MacMaster F. P., Keshavan, M. S., Fitzgerald, K. D., Stewart, C. M. & More, G. J. (2000) Decrease in caudate glutamatergic concentrations in pediatric obsessive-compulsive disorder patients taking paroxetine. *J Am Acad Child Adolesc Psychiatry.*, *39*, 1096 – 103.

[56] Bolton, J., Moore, G. J., MacMillan, S., Stewart, C. M. & Rosenberg, D. R. (2001) Case study: caudate glutamatergic changes with paroxetine persist after medication discontinuation in pediatric OCD. *J Acam Child Adolesc Psychiatry.*, *40*, 903 -6.

[57] Stein, D. J. (2002). Obsessive-compulsive disorder. *Lancet.*, *360*, 397 – 405.

[58] Kellner, M. (2010). Drug treatment of obsessive compulsive disorder. *Dialogues Clin Neurosci.*, *12*, 187 – 97.

[59] Bokor, G. & Anderson, P. D. (2014). Ketamine: an update on its abuse. *J Pharm Pract.*, *27*, 582 – 6.

[60] Rodriguez, C. I., Kegeles L. S., Flood, P. & Simpson, H. B. (2011). Rapid resolution of obsessions after an infusion of intravenous ketamine in a patient with treatment-resistant obsessive-compulsive disorder. *J Clin Psychiatry.*, *72*, 567 – 9.

[61] Bloch, M. H., Wasylink, S., Landeros-Weisenberger, A., Panza, K. E., Billingslea, E. & Leckman, J. F. et al. (2012). Effects of ketamine in treatment-refractory obsessive-compulsive disorder. *Biol Psychiatry.*, *72*, 964 – 0.

[62] Porter, K. (2004). Ketamine in prehospital care. *Emerg Med J.*, *21*, 351 – 4.

[63] Rodriguez, C. I., Kegeles, L. S., Levinson, A., Fend, T., Marcus, S. M., Vermes, D. et al. (2013). Randomized controlled crossover trial of ketamine in obsessive-compulsive disorder: proof-of-concept. *Neuropsychopharmacology.*, *38*, 2475 – 83.

[64] Moghaddam, B., Adams, B., Verma, A. & Daly, D. (1997). Activation of glutamatergic neurotransmission: a novel step in the pathway from NMDA receptor blockade to dopaminergic and cognitive disruptions associated with the prefrontal cortex. *J Neurosci.*, 17, 2921 – 7.

[65] Iadarola, N. D., Niciu, M. J., Richards, E. M., Vande Voort, J. L., Ballard, E. D., Lundin, N. B. et al. (2015). Ketamine and other N-methyl-D-aspartate receptor antagonists in the treatment of depression: a perspective review. *Ther Adv Chronic Dis.*, *6*, 97 – 114.

[66] Nagoshi, N., Nakashima, H. & Fehlings, M. G. (2015). Riluzole as a neuroprotective drug for spinal cord injury: from bench to bedside. *Molecules.*, *20*, 7775 – 89.

[67] Coric, V., Taskiran, S., Pittenger, C., Wasylink, S., Mathalon, D. H., Valentine, G. et al. (2005). Riluzole augmentation in treatment-resistant obsessive-compulsive disorder: an open-label trial. *Biol Psychiatry.*, 58, 424 -8.

[68] Grant, P., Lougee, L., Hirschtritt, M. & Swedo, S. E. (2007). An open-label trial of riluzole, a glutamate antagonist, in children with treatment-resistant obsessive-compulsive disorder. *J Child Adolesc Psychopharmacol.*, *17*, 761 – 7.

[69] Grant, P. J., Joseph, L. A., Farmer, C. A., Luckenbaugh, D. A., Lougee, L. C., Zarate, C. A. & Swedo, S. E. (2014). 12-week, placebo-controlled trial of add-on riluzole in the treatment of childhood-onset obsessive-compulsive disorder. *Neuropsychopharmacology.*, *39*, 1453 – 9.

[70] Olivares, D., Deshpande, V. K., Shi, Y., Lahiri, D. K., Greig, N. H., Rogers, J. T. & Huang, X. (2012). N-methyl-D-aspartate (NMDA) receptor antagonists and memantine treatment for Alzheimer's disease, vascular dementia and Parkinson's Disease. *Curr Alzheimer Res.*, *9*, 746 -58.

[71] Rammes, G., Rupprecht, R., Ferrari, U., Zieglgansberger, W. & Parsons, C. G. (2001). The N-methyl-D-aspartate receptor channel blockers memantine, MRZ 2/579 and other amino-alkyl-cyclohexanes antagonize 5-HT(3) receptor currents in cultured HEK-293 and N1E-115 cell systems in non-competitive manner. *Neurosci Lett.*, *306*, 81-4.

[72] Seeman, P., Caruso, C. & Lasaga, M. (2008). Memantine agonist action at dopamine D2 high receptors. *Synapse.*, *62*, 149 – 53.

[73] Stewart, S. E., Jenike, E. A., Hezel, D. M., Stack, D. E., Dodman, N. H., Shuster, L. & Jenike, M. A. (2010). A single-blinded case-control study of memantine in the severe obsessive-compulsive disorder. *J Clin Psychopharmacol.*, *30*, 34 – 9.

[74] Ghaleiha, A., Entezari, N., Modabbernia, A., Najand, B., Askari, N., Tabrizi, M. et al., (2013). Memantine add-on in moderate to severe obsessive-compulsive disorder: randomized double-blind placebo-controlled study. *J Psych Res.*, *47*, 175 -80.

[75] Hadhighi, M., Jahangard, L., Mohammad-Beigi, H., Bajoghli, H., Hafezian, H., Rahmimi, A. et al. (2013). In a double-blind, randomized and placebo-controlled trial, adjuvant memantine improved symptoms in inpatients suffering from refractory obsessive-compulsive disorder (OCD). *Psychopharmacology.*, *228*, 633 – 40.

[76] Dean, O., Giorlando, F. & Berk, M. (2011). N-acetylcysteine in psychiatry: current therapeutic evidence and potential mechanisms of action. *J Psychiatry Neurosci.*, *36*, 78 – 6.

[77] Oliver, G., Dean, O., Camfield, D., Blair-West, S., Ng, C., Berk, M. & Sarris, J. (2015). N-acetyl cysteine in the treatment of obsessive compulsive and related disorders: a systematic review. *Clin Psychopharmcol and Neurosci.*, *13*, 12 -24.

[78] Moran, M. M., McFarland, K., Melendez, R. I., Kalivas, P. W. & Seamans, J. K. (2005). Cystine/glutamate exchange regulates metabotropic glutamate receptor presynaptic inhibition of excitatory transmission and vulnerability to cocaine seeking. *J Neurosci.*, *25*, 6389 – 93.

[79] Fritzen, F. M. & Bloch, M. H. (2014). N-acetylcysteine for the treatment of obsessive-compulsive disorder. *Curr Psychiatry Rev.*, *10*, 308-316.

[80] Lafleur, D. L., Pittenger, C., Kelmendi, B., Gardner, T., Wasylink, S. & Malison, R. T. (2006). N-acetylcysteine augmentation in serotonin reuptake inhibitor refractory obsessive-compulsive disorder. *Psychopharmacology.*, *184*, 254 -56.

[81] Egashira, N., Shirakawa A., Abe M., Niki T., Mishima K., Iwasaki K. et al. (2012). N-acetyl-l-cysteine inhibits marble-burying behavior in mice. *J Pharmacol Sci.*, *119*, 97 – 01.

[82] Afshar H., Roohafza, H., Mohammad-Beigi, H., Haghighi, M., Jahangard, L., Shokouth, P. et al. (2012). N-acetylcysteine add-on treatment in refractory obsessive-compulsive disorder. *J Clin Psychopharmacol.*, *32*, 797 – 03.

[83] Grant, J. E., Odlaug, B. L. & Kim, S. W. (2009). N-acetylcysteine, a glutamate modulator, in the treatment of trichotillomania: a double-blind, placebo-controlled study. *Arch Gen Psychiatry.*, *66*, 756 – 63.

[84] Chasson, G. S., Buhlman, U., Tolin, D. F., Sowmya, R. R., Reese, H. E., Rowley, T. et al. (2010). Need for speed: evaluating slopes of OCD recovery in behavior therapy enhanced with D-cycloserine. *Behav Res and Ther.*, *48*, 675-679.

[85] Storch, E. A., Murphy, T. K., Goodman, W. K., Geffken, G. R., Lewin, A. B., Henin A. et al. (2010). A preliminary study of D-cycloserine augmentation of cognitive-behavioral therapy in pediatric obsessive-compulsive disorder. *Biol Psychiatry.*, *68*, 073-1076.

[86] Xia, J., Du, Y., Han, J., Liu, G. & Wang, X. (2015). D-cycloserine augmentation in behavioral therapy for obsessive-compulsive disorder: a meta-analysis. *Drug Des Devel Ther.*, *9*, 2101-17.

[87] Andersson, E., Hedman, E., Enander, J., Djurfedt, D. R., Ljotsson, B., Cervenka, S. et al. (2015). D-cycloserine vs placebo as adjunct to cognitive behavioral therapy for obsessive-compulsive disorder and interaction with antidepressants. *JAMA Psychiatry.* Doi:10.1001/ jamapsychiatry.2015.0546.

[88] Skolnick, P., Layer, R. T., Popik, P., Nowak, G., Paul, I. A. & Trullas, R. (1996). Adaptation of N-methyl-D-aspartate (NMDA) receptors following antidepressant treatment: implications for the pharmacotherapy of depression. *Pharmacopsychiatry.*, *29*, 23 – 6.

[89] Bruno, A., Micu, U., Pandolfo, G., Mallamace, D., Abenavoli, E., Di Nardo, F. et al. (2012). Lamotrigine augmentation of serotonin reuptake inhibitors in treatment resistant obsessive-compulsive disorder: a double-blind, placebo-controlled study. *J Psychopharmacol.*, *26*, 1456 – 62.

[90] Hussain, A., Dar, M. A., Wani, R. A., Shah. M. S., Jan, M. M., Malik, Y. A. et al. (2015). Role of lamotrigine augmentation in treatment-resistant obsessive compulsive disorder: a retrospective case review from South Asia. *Indian J Psychol Med.*, *37*, 154 – 8.

In: Modulators of Glutamatergic Signaling
Editor: Zoran M. Pavlovic
ISBN: 978-1-63483-493-3

Chapter 6

THE GLUTAMATERGIC SYSTEM IN ALZHEIMER'S DISEASE PATHOLOGY

***S. Kar*[1,2,*] *and G. B. Baker*[2]**

[1]Department of Medicine (Division of Neurology),
Centre for Prions and Protein Folding Diseases,
University of Alberta, Edmonton, Alberta, Canada
[2]Neurochemical Research Unit, Department of Psychiatry
University of Alberta, Edmonton, Alberta, Canada

ABSTRACT

Glutamate, the major excitatory neurotransmitter of the brain, is involved in a variety of functions including synaptic transmission, neuronal growth and differentiation as well as regulation of memory formation. Excess levels of this neurotransmitter have long been suggested to participate in the degeneration of neurons *via* excitotoxicity in Alzheimer's disease (AD), the most prevalent cause of dementia affecting elderly people in our society. Neuropathologically, AD is characterized by the presence of extracellular neuritic plaques containing amyloid β (Aβ) peptide, intracellular neurofibrillary tangles containing tau protein and loss of neurons in selected brain regions including hippocampus and cortex – the regions involved in learning and memory processing. This chapter provides an overview of the altered glutamatergic system in AD pathology, and how its interactions with Aβ peptides can influence synaptic function, tau phosphorylation and neurodegeneration. In fact, glutamate not only influences Aβ production, but Aβ peptide can also alter the levels of glutamate at the synapse, which may influence development and progression of AD pathology. Finally, we discuss the effectiveness of the glutamate receptor antagonist, memantine, in the treatment of patients with AD.

Keywords: amyloid precursor protein, β-amyloid peptides, excitotoxicity, glutamate receptors, memantine, synaptic plasticity

* Corresponding author: S. Kar PhD, Centre for Prions and Protein Folding Diseases, Departments of Medicine and Psychiatry, University of Alberta, Edmonton, Alberta, Canada T6G 2M8; email: skar@ualberta.ca.

INTRODUCTION

Alzheimer's disease (AD), described first in 1906 by the German physician Alois Alzheimer, is currently considered to be the most common cause of dementia, accounting for about 50-60% of all age-related dementia in individuals over 65 yrs of age [1-3]. It is a progressive disease characterized by a global deterioration of intellectual function that includes an amnesic type of memory impairment, deterioration of language and visuospatial deficits. Motor and sensory abnormalities are uncommon until the late phases of the disease, and basic activities of daily living are gradually impaired as the disease enters advanced phases. Psychosis and agitation also develop during middle or later phases of the disease. The average course of AD is approximately a decade, but the rate of progression is quite variable from individual to individual [4-6]. At present, there is no remission in the progression of the disease nor is there any truly effective treatment. Epidemiological data have shown that AD afflicts about 8-10% of the population over 65 years of age and its prevalence doubles every 5 years thereafter. With the proportion of elderly people increasing steadily in our population, AD is expected to pose an increasing economic challenge to our society over the coming decades [3, 6].

While AD aetiology still remains unclear, evidence gathered over the years suggests that both genetic and environmental factors contribute to disease pathogenesis and progression. In the majority of cases AD appears to occur as sporadic disease after the age of 65 years, but in a small proportion of cases the disease is inherited as an autosomal dominant trait and appears as an early onset form prior to 65 years of age. To date, mutations in three genes – the amyloid precursor protein (*APP*) gene on chromosome 21, the presenilin 1 (*PSEN1*) gene on chromosome 14 and the presenilin 2 (*PSEN2*) gene on chromosome 1 – have been identified as the cause of early onset familial AD [5, 7, 8]. While these findings are of importance in elucidating the biological pathogenesis of AD, it is vital to recognize that mutations in these three genes may only account for 30-50% of all autosomal dominant early-onset cases. Although no genetic causes of late-onset AD have yet been identified, association studies have indicated that certain candidate genes can increase the risk of developing late-onset AD. One such genetic risk factor is possession of one or two copies of the ε4 allele of the gene for apolipoprotein, which has been shown to increase the probability of developing AD [9, 10]. A number of genome-wide association studies have shown that variants in some other genes such as clusterin (*APOJ* on chromosome 8) [11, 12], phosphatidylinositol-binding clathrin assembly protein (*PICALM* on chromosome 11) [11, 12], complement receptor 1 (*CR1* on chromosome 1; [12]), bridging integrator 1 (*BIN1* on chromosome 2) [13] and sortilin-related receptor1 (*SORL1* on chromosome 11) [14] may enhance the risk of AD, but these need to be validated [15, 16]. Other factors that play an important role in AD pathogenesis include age, head injury and oxidative stress [17, 18].

NEUROPATHOLOGICAL FEATURES OF AD

Neuropathologically AD is characterized by three distinct features -- intracellular neurofibrillary tangles (NFTs), extracellular neuritic plaques and neuronal loss [3, 19]. Neurofibrillary tangles are intracellular aggregates of the hyperphosphorylated microtubule associated protein tau, which plays a critical role in the stabilization of microtubules in the normal brain. It is suggested that hyperphosphorylation of tau protein may trigger death of affected neurons by disrupting neuronal transport [20-22]. Tangles are abundant in the entorhinal cortex, hippocampus, amygdala, association cortices of the frontal, temporal and parietal lobes, and certain subcortical nuclei that project into these areas [20-24]. The number and distribution of NFTs in the cortical region of the brain correlate positively with cognitive deficits and serve as a good marker of disease progression [25]. Extracellular neuritic plaques, on the other hand, are multicellular lesions containing a central deposit of amyloid peptides surrounded by reactive astrocytes, activated microglia and dystrophic neurites. The major amyloid peptides that are found in the plaques are β-amyloid$_{1\text{-}42}$ (Aβ$_{1\text{-}42}$) and Aβ$_{1\text{-}40}$, peptides generated by proteolytic cleavage of APP. These plaques are mostly evident in the entorhinal cortex, hippocampus and association cortices, the primary areas affected by neurodegeneration [3, 26-28]. The number of plaques does not correlate with the severity of dementia, but a clinical correlation between elevated levels of Aβ peptide in the brain and cognitive decline has been reported [29]. Several lines of evidence suggest that overproduction and/or reduced clearance of Aβ peptides are likely key to amyloid aggregation, which in turn contributes to the development of NFTs and neurodegeneration in AD [3, 27, 30, 31]. Regions that are preferentially affected in AD brains include the hippocampus, entorhinal cortex, amygdala, neocortex, some subcortical areas such as basal forebrain (cholinergic neurons), the dorsal raphe (serotonergic neurons) and the locus coeruleus (noradrenergic neurons) [3, 4, 32, 33]. There is evidence that glutamatergic neurons located in the hippocampus as well as in frontal, temporal and parietal cortex are severely affected, whereas similar neurons in the motor and sensory cortex are relatively spared [34, 35]. A recent study also revealed that glutamatergic pre-synaptic bouton density increases in patients with mild cognitive impairment (MCI) but decreases in mild and severe AD along with observed cognitive deficits, suggesting that there may be an initial compensatory event in MCI brains which decays with the progression of disease pathology [36, 37]. At present, the underlying cause of neurodegeneration in AD brains has not been clearly established. However, an increasing amount of evidence derived from genetic, epidemiological and biochemical studies has suggested a critical role for Aβ peptides in AD pathogenesis. It is believed that an interplay between Aβ peptides and glutamatergic transmission by influencing multiple pathways may render the neurons vulnerable, leading to the development of AD pathology [38-40]. This review is intended to give an overview of the interactions between the central glutamate system and Aβ peptides and their relevance in the development of AD pathology.

Glutamatergic System in the AD Brain

Glutamate, the major excitatory neurotransmitter, is found in more than 80% of all neurons in the central nervous system. It is considered to play a critical role in synaptic transmission, neuronal growth and differentiation, synaptic plasticity and learning and memory [34, 35]. Glutamatergic neurons are located in all major brain areas including the cortex and hippocampus – the regions severely affected in AD pathology. In pre-synaptic neurons, glutamate is derived primarily from glucose and α-ketoglutarate and to some extent from glutamine transported from glial cells. Following synthesis, glutamate is packaged into vesicles by the vesicular glutamate transporters, VGLUT1 and VGLUT2, which maintain the level of glutamate in secretory vesicles [41]. When a neuron is depolarised, glutamate is released into the synaptic cleft where it binds glutamate receptors on pre- and post-synaptic neurons to mediate synaptic transmission [42-45]. The excess glutamate is cleared from the synaptic sites mostly by astrocytes which express high levels of the glutamate transporters GLAST and GLT-1 in the rat [46-48]. The human homologues are the excitatory amino acid transporters 1 and 2 (EAAT1 and 2) [46, 49; Figure 1]. In astrocytes glutamate is converted to glutamine by the enzyme glutamine synthetase and then transported back to neurons for utilization in glutamate synthesis. Glutamatergic transmission is mediated by two families of receptors: ionotropic (iGluR) and metabotropic (mGluR) glutamate receptors. The iGluR family is divided into three subtypes: N-methyl-D-aspartate (NMDA containing: GluN1, GluN2A-D and GluN3A-B subunits) receptors permeable to Ca^{2+} ions, whereas α-amino-3-hydroxy-5-methyl-4-isoxazolepropionic acid (AMPA containing GluA1-4 subunits) and kainate (KA containing GluK1-5 subunits) receptors are permeable to Na^{+} and K^{+} ions [42, 43, 45]. Activation of NMDA receptors by glutamate/NMDA and co-agonist (glycine/D-serine) requires membrane depolarization to remove the Mg^{2+} ion block of the channel pore that occurs at resting membrane potentials [50]. A number of studies have indicated that NMDA receptors, depending on the subunit composition, can be located either on the synaptic or extrasynaptic sites and are coupled to different cellular events. In general, synaptic NMDA receptors with GluN1/GluN2A subunits, which require glycine as co-agonist, activate pro-survival signaling, while extrasynaptic NMDA receptors containing GluN1/GluN2B subunits are usually gated by the co-agonist D-serine and are known to be associated with pro-death signaling pathways [45, 50, 51]. As for metabotropic glutamate receptor family, there are eight different metabotropic G-protein coupled receptors, which on the basis of pharmacology, intracellular cascade and sequence homology are classified into three groups: Group I (mGluR1 and mGluR5), Group II (mGluR2 and mGluR3) and Group III (mGluR4 and mGluR6-8) receptors [42, 52]. Unlike iGluRs which are primarily situated in the post-synaptic membrane to mediate fast transmission, mGluRs are located in various membrane compartments of both neuronal and glial cells [52, 53]. The only mGluR not found in the brain is mGluR6, which is expressed in the retina [52, 54].

Deficiencies in many stages of the glutamate cycle have been shown to occur in AD, leading to elevated baseline levels of glutamate around the neurons of the synapse. The glutamatergic neurons are situated in areas that are known to be affected in AD. The pyramidal neurons of the neocortex [55, 56] as well as neurons located in the entorhinal cortex and hippocampal CA1 region [35] are lost early in disease progression, whereas some of the surviving neurons are susceptible to the formation of NFTs, highlighting their significance in AD pathogenesis [34, 35]. Supporting the histological data, some biochemical studies have shown decreased concentrations of glutamate *in vivo* as well as in post-mortem brain tissue [35, 57]. There have been a number of studies that noted reduced levels of VGLUT1 and 2 in the prefrontal cortex of AD, with VGLUT1 showing a much greater reduction than VGLUT2 [57, 58]. More recent studies using synaptosome-enriched fractions also showed a trend for reduced VGLUT1 and VGLUT2 levels in the parietal cortex of AD brains compared to controls, but they did not reach significant levels [59]. The gene expression and protein levels of EAAT1 and EAAT2 have also been shown to be reduced in the cortex and hippocampus of AD patients [59-62], suggesting that glutamate clearance from the synaptic site is attenuated. Changes in glutamate transporter expression have been identified at an early stage of AD, indicating that dysfunction in glutamate transport might be an early event in the disease pathology [62, 63]. There is also a reduction in glutamine synthetase in AD [64], thus preventing the transportation of glutamate back to pre-synaptic neurons. Collectively, these changes can lead to increased concentrations of glutamate at the synapse, which can trigger excitotoxic cell death *via* NMDA receptor-mediated increases of intracellular Ca^{2+} [65-67]. Elevated Ca^{2+} levels lead to neuronal death through a number of pathways, including increased activation of calpain [68], release of reactive oxygen species, and depolarisation of the mitochondrial membrane, which leads to reduced energy metabolism and cytochrome c release [65]. These ultimately result in neuronal loss through apoptotic and necrotic pathways [65, 67; Figure 1].

Alterations in the levels of iGluRs remain inconclusive. Some reports suggest that there is an increase in NMDA-sensitive glutamate binding in the striatum of AD brains [69], whereas other studies indicate that there are no changes in the level of NMDA receptors in the affected areas of the AD brain [70]. The majority of reports, however, suggest a decrease in protein and mRNA levels of these receptors, at least in later stages of the disease [71-74]. This is supported by autoradiography studies which showed a reduction in NMDA receptor binding in the CA1 and CA3 regions of the hippocampus, supporting the idea that there are fewer NMDA receptors in these areas of the brain. Western blot analysis and immunohistochemical evidence have also shown changes in the levels of AMPA receptors. These results suggest a reduction in GluA1 and GluA2/3 levels in early AD brains, whereas GluA2/3 levels are increased in severe AD cases [75]. Detailed studies of AD brains have shown that the reductions in AMPA receptors are well correlated to the affected areas of the brain, but there is significant variation in AMPA receptor levels between AD brains [76, 77]. These researchers showed GluA1 and GluA2/3 loss was restricted to the subiculum and CA1 layer of the hippocampus, which indicates that changes in AMPA receptor levels are likely to be region specific. The KA receptors have received less attention compared to other classes of iGluRs in AD pathology. Autoradiography studies suggest little change in KA receptor binding in unaffected areas, but a reduction in hippocampus of AD brains compared to controls [78]. Immunohistochemical staining using an antibody directed against GluA5/6/7 also showed some reduction in the CA1 region of the hippocampus of AD brains, which may partly be due to loss of neurons rather than alterations in receptor levels [79]. As for mGluRs, radioligand binding assays showed a reduction in the levels of mGluRs in the cortex and hippocampus of AD brains [78]. Western blot analysis confirms that there is a reduction in mGluR1 levels in CA1 and subiculum of the hippocampus, but other regions that are less susceptible to damage in AD, such as CA3 region, are unaffected [78, 80]. Conversely, increased levels of mGluR2 are found in the hippocampus of AD brains [78, 81]. It is therefore probable that alterations in mGluR expression are involved in AD pathogenesis. Over the last decade substantial new information has been generated with regard to role of the glutamatergic system in AD; these findings have nicely been documented by several investigators recently [38-40, 50, 51, 65, 67]. This review is intended to provide an overview of the interactions between the glutamatergic system and Aβ and their relevance in the development of AD pathology.

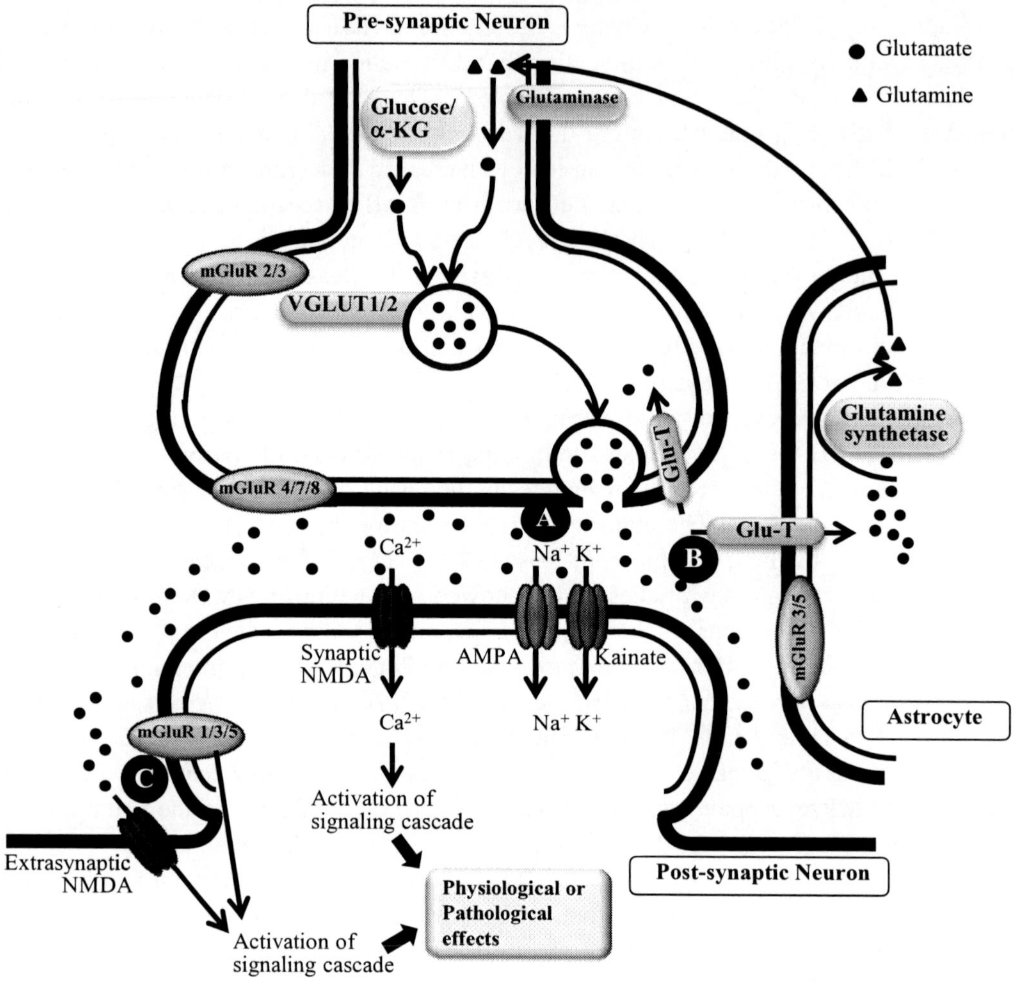

Figure 1. Glutamate, the most abundant excitatory neurotransmitter in the brain, is primarily derived from two sources i) de novo synthesis through glucose and α-ketoglutarate, and ii) from glutamine transported from glial cells. Once synthesized, glutamate is transported to the synaptic vesicles by vesicular glutamate transporters (VGLUT1/2). Following neuronal depolarization, glutamate is released into the synaptic cleft and activates two families of receptors; ionotropic (NMDA, AMPA and kainate) and metabotropic (mGluR) receptors. NMDA receptors can become fully active only after synchronised AMPA receptor activation and depolarization of the post-synaptic membrane. Activated glutamate receptors stimulate downstream intracellular signaling pathways in the postsynaptic neurons. Excess glutamate from the synaptic cleft is removed via a reuptake mechanism into neurons/astrocytes by another group of glutamate transporters (Glu-T). Within astrocytes, glutamate is converted to glutamine by glutamine synthetase and then returned to the neurons where glutaminase converts glutamine back to glutamate. Many aspects of the glutamate cycle are affected in AD brains, implicating glutamate in disease progression. Aβ-related peptides have been shown to increase glutamate release (A), inhibit clearance of glutamate by astrocytes (B), and affect glutamate receptor activity (C), leading to altered synaptic plasticity (i.e., LTP and LTD) and degeneration of neurons.

GLUTAMATE AND GENERATION OF Aβ PEPTIDES

The glutamatergic system has been shown to regulate Aβ production from its precursor APP, a single transmembrane glycoprotein known to be processed proteolytically *via* two alternative pathways. The majority of APP, following transport to the cell surface, is cleaved by α-secretase enzyme within the Aβ domain, yielding soluble APPα and a 10-kDa C-terminal fragment (CTF-α), which can then be processed by γ-secretase to generate $A\beta_{17-40}/A\beta_{17-42}$. The β-secretase, on the other hand, cleaves APP to generate soluble APPβ and an Aβ-containing C-terminal fragment (CTF-β), which is further processed *via* γ-secretase to yield full-length $A\beta_{1-40}/A\beta_{1-42}$ peptides. While α-secretase processing occurs mostly in the secretory pathway, the endosomal-lysosomal system plays a critical role in the production of Aβ peptides. The candidates that can act as α-secretases are members of the ADAM (a disintegrin and metalloproteinase domain-containing peptidases such as ADAM-9, ADAM-10 or ADAM-17). The β-secretase, on the other hand, is an aspartyl protease called β-site APP cleaving enzyme (BACE1), while the γ-secretase comprises the aspartyl protease presenilin 1/2 and three cofactors (i.e., nicastrin, PEN-2 and APH-1) [3, 27, 81, 82]. In the normal brain Aβ peptides are present in pM-nM concentrations, with ~90% $A\beta_{1-40}$ and ~10% $A\beta_{1-42}$ isoforms. The $A\beta_{1-42}$ isoform is more susceptible to fibril formation as it exhibits higher hydrophobic properties than the $A\beta_{1-40}$ isoform [3, 84]. The physiological functions of these peptides remain poorly understood but they are known to be involved in the regulation of normal and activity-dependent functioning of the brain. Relative activities of the two alternative APP processing pathways can be influenced by a variety of factors, including stimulation of receptors for acetylcholine, serotonin, glutamate, neuropeptides and growth factors [83, 85, 86]. The role of glutamate in APP processing has received a lot of attention as activation of both iGluRs and mGluRs is known to differentially regulate the non-amyloidogenic and amyloidogenic pathways [38, 85].

Activation of NMDA receptors has been shown to either increase or decrease Aβ production depending on the temporal and spatial profile of receptor-evoked intracellular signaling events. Short-term stimulation of synaptic NMDA receptors by agonists has been shown to decrease Aβ production/release *via* a non-amyloidogenic α-secretase pathway. This is triggered first by increased trafficking/activation of ADAM-10 at the cell membrane followed by up-regulation of ADAM-10 expression by an extracellular signal-related kinase (ERK)-dependent signaling mechanism [87, 88]. Conversely, prolonged stimulation of NMDA receptors with sub-maximal doses of agonist or targeted activation of extrasynaptic, but not synaptic, NMDA receptors has been shown to increase the production of Aβ peptides. This effect is preceded by a shift in the expression of APP695 (the neuronal isoform of APP) to Kunitz protease inhibitory domain containing APP751/APP770 isoforms which exhibit greater potential of processing *via* an amyloidogenic β-secretase pathway. This shift is mediated by activation of extrasynaptic NMDA receptors through Ca^{2+}/calmodulin-dependent protein kinase (CaMK) IV [89, 90]. Using *in vivo* microdialysis in living mice, Verges et al. demonstrated that low doses of NMDA can elevate, whereas high doses can suppress, Aβ levels in the interstitial fluid by regulating APP processing. Furthermore, NMDA receptor antagonists can increase interstitial Aβ levels, thus indicating that basal activity of this receptor possibly suppresses Aβ production [91]. Direct stimulation of AMPA receptors has recently been reported to inhibit Aβ production *via* non-amyloidogenic α-secretase-mediated APP processing [92]. Activation of presynaptic group II mGluRs, on the other hand, increases Aβ production/secretion, whereas post-synaptic group I mGluRs promote non-amyloidogenic processing of APP [93, 94]. Collectively, these results suggest that deregulation of glutamatergic transmission can directly influence Aβ production/secretion by differentially regulating APP processing.

GLUTAMATE AND Aβ PEPTIDE IN SYNAPTIC ACTIVITY

Synaptic plasticity is the process by which synapses modulate their strength and form new connections or lose established connections with other neurons. Glutamate is essential in regulating synaptic plasticity that underlies learning and memory through two well characterized phenomena called long term potentiation (LTP) and long term depression (LTD). LTP is generated by high frequency stimulation of the pre-synaptic plasma membrane, resulting in increased release of glutamate and activation of its receptors at the postsynaptic membrane. AMPA and mGlu receptors are activated by the initial glutamate release, whereas GluN2A-containing NMDA receptors only become fully active after continuous synchronised glutamate stimulation following activation of the AMPA and mGlu receptors. NMDA receptor activation allows Ca^{2+} to enter the post-synaptic cell which subsequently triggers different kinase pathways involving CaMKII, ERK and cyclic AMP response element binding protein (CREB). Activation of various kinase pathways increases protein transcription which strengthens synapses and increases synaptic density [39, 95, 96]. LTD, which reduces synaptic strength and prevents memory formation, occurs when there is little stimulation at an established synapse or by asynchronous stimulation of iGluRs. LTD can be caused by a reduction in the levels of NMDA and AMPA receptors at the postsynaptic membrane or by repeated weak stimulation of AMPA receptors that does not lead to the depolarisation of the postsynaptic membrane. This leads to neuronal remodeling with spine shrinkage or reduction in their number and reduced synaptic activity. Multiple lines of evidence suggest that a decrease in LTP and an increase in LTD may underlie the synapse losses and cognitive deficits observed in AD pathology [38, 95, 97].

Using multiple paradigms, it has been shown that Aβ peptides derived from different sources can disrupt LTP and reduce synaptic plasticity [38, 97, 98, 99; Table-1]. In fact, naturally occurring Aβ peptides obtained from spinal fluid [100] or directly from the brains of AD patients [101], and naturally secreted Aβ oligos from cultured cells can inhibit the induction of LTP and synaptic density [102, 103]. Additionally, it has been suggested that Aβ dimers and larger aggregates are required for Aβ-induced inhibition of LTP and that monomeric Aβ is not inhibitory [100, 103]. The inhibitory effects of Aβ on LTP involve multiple kinases, TNF-α and CREB protein and can be reversed by either removing Aβ with monoclonal antibodies or by treatment with the NMDA receptor antagonist memantine as well as the GluN2B negative allosteric modulators ifenprodil and Ro 25-6981 [38, 100, 104]. Thus, activation of GluN2B-containing NMDA receptors located extra-synaptically may mediate Aβ-induced LTP inhibition. Apart from inhibiting LTP, Aβ peptide has been shown to facilitate LTD in an NMDA receptor-dependent manner under both *in vitro* and *in vivo* conditions [97, 102; Table-1]. Soluble Aβ oligomers from several sources (synthetic, cell cultures, human brain extracts) have been shown to facilitate NMDA- and mGluR-dependent LTD in hippocampal slices [38, 105]. Intracerebroventricular injection of Aβ peptide triggers a long-lasting depression of hippocampal excitatory postsynaptic currents that can be averted by an NMDA receptor antagonist [106]. More recently, it has been reported that Aβ-induced depression of glutamatergic transmission, which shares a similar mechanism with NMDA receptor-mediated LTD, can be blocked by the GluN2B specific antagonists ifenprodil and Ro 25-6981 but not by antagonists that block the NMDA receptor channel pore (i.e., MK-801 or ketamine) or the glycine binding site on the GluN1 subunit (i.e., 7-chlorokynurenate) [107, 108]. These findings imply that conformational changes in GluN2B subunits but not calcium influx through the NMDA receptor are possibly involved in mediating the effects of Aβ peptides. Additionally, high levels of soluble Aβ have also been shown to reduce the number and alter the morphology of dendritic spines in mutant APP transgenic mice as evident following treatments of hippocampal slices and primary neurons with exogenous Aβ peptides [97, 103, 109, 110]. Spine density and morphological changes are reversed when cultures are treated with a γ-secretase inhibitor, indicating that Aβ is responsible for reduced synaptic density. Also, NMDA receptor antagonists such as AP-5 prevent Aβ-induced spine loss, suggesting that glutamate is essential not only for LTP/LTD but also for Aβ-induced changes in spine density/morphology [38, 109].

Table 1. The roles of Aβ peptides in regulating glutamatergic neurons

Physiological and Pathological concentrations of Aβ	References
Inhibit long-term potentiation (LTP)	[98, 102, 103, 191]
Increase long-term depression (LTD)	[97, 103, 105]
Decrease synaptic density and alter the morphology of dendritic spines	[97, 109, 110]
Regulate glutamate uptake from the synapse	[61, 105, 134-137]
Regulate glutamate release	[123, 125, 127-130, 153]
Increase tau phosphorylation/cell death	[145-149]

Synaptic transmission, apart from synaptic plasticity, has been shown to be influenced by Aβ peptides. Evidence suggests that increased synaptic activity enhances Aβ production both in *in vitro* and *in vivo* [94, 103, 111], whereas enhanced levels of soluble Aβ in cells over-expressing APP depress synaptic activity [112]. Thus, it is likely that Aβ may act as a regulator of synaptic activity through a negative feedback loop. One possible mechanism by which excess Aβ could reduce synaptic transmission is by increasing the levels and activity of striatal-enriched tyrosine phosphatase ($STEP_{61}$), which has been shown to induce endocytosis of NMDA and AMPA receptors from post-synaptic membranes and reduce synaptic function [113, 114-116].

Some recent reports also indicate that soluble Aβ peptide can disrupt the post-synaptic density (PSD), an electron-dense thickening of the membrane comprised of a network of proteins including glutamate receptors, adhesion molecules, scaffolding/cytoskeletal proteins and associated signaling molecules. The PSD, involved in maintaining synaptic homeostasis, is regulated by cytoskeletal protein PSD-95 and the master organising proteins Shank and Homer [117, 118]. PSD-95 and Shank are important for ensuring AMPA and NMDA receptors get to the cell surface and stay at the plasma membrane [117, 119] and Homer is important in mGluR1 localisation [120]. Increased levels of Aβ trigger a breakdown of the PSD-95, Shank and Homer clusters and a reduction in levels of NMDA and AMPA receptors at the plasma membrane [118]. Thus, Aβ-induced synaptic dysfunction may partly be mediated by its ability to affect elements of the synaptic machinery, which can influence pathological abnormalities in AD brains (Figure 1).

Glutamatergic Transmission and Aβ Peptide

Although Aβ peptides are produced constitutively by brain cells and are found in the nM - pM range in the interstitial fluid [3, 27], their physiological role in the brain remains unclear. Several studies have suggested that these peptides may regulate the release of brain neurotransmitters/ modulators in the absence of evident signs of neurotoxicity. Indeed, we and others have reported that nM - pM concentrations of $A\beta_{1\text{-}40/1\text{-}42}$ can inhibit the release of acetylcholine, particularly from hippocampal and cortical regions of the rat brain [33, 121, 122]. Aβ-related peptides can also directly modulate the release of noradrenaline [123, 124], but not of GABA [125], from rat brain slice preparations. There is evidence that Aβ peptides can disrupt the cholinergic control of dopamine release but do not affect either basal or potassium-evoked dopamine release from nucleus accumbens or caudate putamen of the brain [126]. A study by Arias et al. (1995) demonstrated that pre-treatment of hippocampal slices with μM concentrations of $A\beta_{25\text{-}35}$ can potentiate glutamate release [125]. Using an *in vitro* slice preparation, we have shown that nM concentrations of $A\beta_{1\text{-}40/1\text{-}42}$ can directly potentiate potassium-stimulated glutamate release from the hippocampus and cortex – the areas primarily affected in AD brains [127]. Aβ-related peptides have also been shown to promote glutamate release from cortical nerve endings [123] and cultured hippocampal neurons [128] and to modulate nicotine receptor-induced glutamate release from adult rat hippocampal synaptosomal preparations [129]. Interestingly, a recent study showed that nM-pM concentrations of oligomeric $A\beta_{1\text{-}42}$ can, by activating nicotinic α_7 nicotinic receptors, potentiate glutamate release from cultured astrocytes [130]. These findings, taken together, suggest a role for endogenous Aβ in the regulation of glutamate release from selected brain regions. Although the underlying mechanisms by which Aβ can potentiate glutamate release remain unclear, it is of interest to note that nM concentrations of $A\beta_{1\text{-}40}$, under acute conditions, did not alter glutamate release from the striatum, an area which is relatively spared in AD brains [127]. Whether prolonged exposure to μM concentrations of Aβ peptide can enhance glutamate release from striatal slices remains to be established. Nevertheless these results raise the possibility that preferential vulnerability of hippocampal and cortical neurons may partly relate to their sensitivity to Aβ. This is supported by recent data which showed that potassium-evoked glutamate release in the CA1 region of the hippocampus was increased in mutant APP/ PS1 transgenic mice preceding the development of neuropathology and cognitive decline [131].

Apart from glutamate release, Aβ peptides have been shown to modulate glutamate uptake *via* neuronal and glial transporters. Some studies have suggested that μM concentrations of Aβ-related peptides can enhance glutamate uptake by astrocytes/neurons which is mediated by increased expression of the glutamate transporter GLAST/GLT1 and enhanced by ERK inhibition [132-134]. Conversely, a number of studies have shown that nM-μM concentrations of Aβ peptides are able to increase extracellular glutamate levels by inhibiting its uptake into cultured neurons and astrocytes [135-138]. It is proposed that oxidative stress induced by Aβ damages glutamate transporters, preventing uptake of glutamate [138]. More recent data have confirmed that $A\beta_{1-40}$, by triggering oxidative damage and decreasing GLT-1 and GLAST levels, can reduce the clearance of glutamate from the synapse [135]. Soluble Aβ isolated from multiple sources (synthetic, cell culture and human brain extracts) has also been found to decrease uptake of glutamate, resulting in GluN2B-containing NMDA receptor-mediated extrasynaptic responses [105]. Prolonged exposure of cultured microglia to 5μM $A\beta_{25-35}$, on the other hand, has been shown to increase extracellular glutamate levels *via* reverse glutamate transport [139]. Thus Aβ peptides can regulate extracellular glutamate levels by multiple mechanisms which may have direct relevance in normal physiological as well as pathological conditions (Table 1).

GLUTAMATE AND Aβ TOXICITY

Chronic exposure to Aβ peptides can induce toxicity in a variety of cell lines, as well as in primary rat and human cultured neurons. Although toxic properties of the peptide relate to its ability to form insoluble aggregates [27, 140-142], recent data suggest that the most detrimental forms of Aβ peptides are the soluble oligomers and that the insoluble amorphous or fibrillar deposits represent a less harmful inactivated form of the peptide [3]. The mechanisms associated with Aβ toxicity are not clearly defined, but appear to involve alterations in intracellular calcium, production of free radicals, phosphorylation of tau protein and/or activation of caspase- and non-caspase pathways that culminate in programmed cell death [27, 143, 144]. A number of earlier studies suggested that Aβ toxicity is induced, at least in part, by glutamate-mediated excitotoxicity – which involves activation of the NMDA receptors, leading to elevated intracellular Ca^{2+} and consequent stimulation of a cascade of enzymes, resulting in cell death [39, 65, 67, 68, 142]. This is supported by data which showed that: i) inactivation of the NMDA receptor by antagonists such as MK-801, AP5 or memantine can partially protect neurons from Aβ1-40/1-42 toxicity [145-148]; ii) Aβ1-40/1-42-induced neurodegeneration in the adult rat brain is mediated, in part, by activation of the NMDA receptor [149, 150] and iii) transgenic mice exhibiting high levels of Aβ peptide show increased vulnerability to excitotoxicity [151, 152].

At present, however, it is unclear as to how Aβ peptides can regulate activation of glutamate receptors, leading to the death of neurons. There is evidence that Aβ peptide can bind directly to glutamate receptors, increasing their activity and leading to influx of Ca^{2+} in organotypic slice cultures. Antagonists of NMDA and AMPA receptors can prevent influx of Ca^{2+} and cell death [39, 65, 84]. Alternatively, it is possible that Aβ-related peptides can increase extracellular glutamate levels by potentiating release [127, 128-130, 139, 153] and/or inhibiting the uptake of the neurotransmitter [136, 137], which can subsequently trigger death of neurons by excitotoxicity. It is therefore likely that a combination of reduced glutamate clearance from the synaptic cleft, increased release of glutamate from neurons and glia, and the subsequent activation of glutamate receptors contribute to toxicity mediated by Aβ-related peptides (Table-1). In fact, a number of recent studies have indicated that Aβ toxicity is partly mediated by activating GluN2B-containing extrasynaptic NMDA receptors coupled to cell death signaling pathways. This is supported by the evidence that i) GluN2B antagonists such as memantine, ifenprodil [154] and Ro 25-6981, but not of GluN2A antagonists, are able to partially protect cells against Aβ-mediated toxicity [155, 156], ii) activation of NMDA receptors containing GluN2B subunit results in neuronal apoptosis under both *in vitro* and *in vivo* paradigms [157] and iii) reduction of D-serine content in serine racemase knockout mice markedly attenuates toxicity induced by NMDA and Aβ peptide [158]. Thus, selective enhancement of GluN2A activity and/or attenuation of GluN2B activity may be useful in preventing Aβ-induced toxicity.

Given the significance of tau phosphorylation that precedes loss of neurons triggered by Aβ-peptide, a functional interrelationship between glutamate receptors and phosphorylation of tau protein has long been investigated in a variety of experimental paradigms. We and others have shown that glutamate receptor antagonists can protect neurons against $A\beta_{1-40/1-42}$-mediated toxicity [146-148]. This effect appears to be partly mediated by reversing/ decreasing activation of various tau-kinases such as ERK1/2, glycogen synthase kinase-3β (GSK-3β) and Fyn kinase that are involved in the tau phosphorylation rather than altering extracellular glutamate levels or regulation of Aβ conformation/internalization [147; Table 1]. There is evidence that Fyn kinase can also regulate glutamatergic NMDA receptor activation following Aβ treatment by triggering phosphorylation and subsequent interaction of the GluN2B subunit of the receptor with PSD-95, an important postsynaptic scaffolding protein that plays a role in protein assembly, synaptic development and neural plasticity. Disrupting the localization of Fyn kinase or the formation of the GluN2B/PSD-95/Fyn complex was found to protect neurons against Aβ-induced toxicity [159, 160]. Collectively, these results suggest that not only increased tau phosphorylation, but also transportation of Fyn kinase to postsynaptic sites, is possibly required for degeneration of neurons induced by Aβ peptides.

Increased phosphorylation of tau protein may also be a consequence of reduced activity of phosphatases. The serine/threonine phosphatases, such as PP-1, PP-2A, PP-2B (i.e., calcineurin) and PP-5 have all been shown to dephosphorylate tau, and the activity of PP-1, PP-2A and PP-5 is known to be reduced in AD brains [161-163].

Since PP-1 and PP-2A account for ~90% of serine/threonine phosphatase activity in mammalian cells [164, 165] and tau dephosphorylation is known to be mediated predominantly by PP-2A [161], it is believed that down regulation of this phosphatase may underlie increased tau phosphorylation in AD brains. Interestingly, inhibiting PP-2A activity has been shown to increase tau hyperphosphorylation by reducing its dephosphorylation [166, 167] as well as by increasing the activity of tau kinases such as CaMKII [166], GSK-3β [168] and ERK1/2 [169]. Thus, reduced activity of phosphatases may also partly contribute to the increased tau phosphorylation and loss of neurons. Some studies have indicated that caspase- and calpain-mediated cleavage of tau protein may also have a role in Aβ-mediated toxicity. There is evidence that 43-50kDa truncated tau generated by caspase-3 assembles more rapidly into filaments than full-length tau and has been detected prior to the formation of NFTs and cell death [170-173]. Similarly, 17 kDa tau generated by calpain following Aβ treatment was also detected prior to tau phosphorylation in cultured neurons. These cleavages of tau protein may lead to neurodegeneration either directly as reported in various cell lines/neurons or indirectly by reducing the pool of full length tau available for binding to microtubules [174, 175].

GLUTAMATE RECEPTOR ANTAGONISM AS A THERAPEUTIC TARGET IN AD

Given the relevance of aberrant glutamatergic transmission in the development of AD, treatment strategies that can alter the activity of remaining glutamatergic neurons, without causing excitotoxicity, represents an important therapeutic target for AD pathology. Some recent studies have shown that restoring glial glutamate transporters EAAT2 [176] or attenuating functioning of mGluR5 receptors [177] may have disease-modifying beneficial effects in AD pathology. Excitotoxicity mediated by NMDA receptors, on the other hand, has long represented an attractive therapeutic target in AD even though this receptor is critical in regulating normal physiological activity such as neuronal plasticity and neurotrophic processes. However, many NMDA receptor antagonists including, MK-801 and phencyclidine, that showed promise as inhibitors of excitotoxicity also blocked normal synaptic functions. Consequently, severe side effects such as hallucinations, ataxia, and memory loss preclude their use as therapeutics for AD [178, 179]. Memantine, which was first synthesized by Eli Lilly and Company in 1968 as a derivative of amantadine (an NMDA receptor antagonist) is an uncompetitive, low to moderate affinity NMDA receptor antagonist that can prevent pathological activation of the receptor without affecting its physiological functions [180-183]. This antagonist exhibits a lower binding affinity than MK-801 and phencyclidine, but has a higher affinity than the endogenous NMDA receptor antagonist, the magnesium ion, which normally blocks the voltage-dependant activation of the NMDA receptor. Further studies have indicated that memantine preferentially blocks NMDA receptor activity when the channel is excessively open. This particular phenomenon makes memantine a potential neuroprotective agent against excitotoxicity without interfering with normal synaptic activity [181, 182, 184]. More recently, memantine has been demonstrated to selectively target glutamatergic GluN2B-containing NMDA receptors which are located in extrasynaptic sites and are linked to cell death mechanisms [185, 186]. Over the years, memantine has been shown to provide protection against a variety of insults including quinolinic acid-induced toxicity, brain trauma, ischemia and $A\beta_{1-40/1-42}$-mediated neurotoxicity under both *in vitro* and *in vivo* conditions [146-148, 150, 187; Table 2]. The protective effects of memantine have been shown to be increased by coadministration with galantamine – an acetylcholinesterase inhibitor approved for the treatment of AD patients [188, 189]. Interestingly, memantine has been found to protect both glutamatergic as well as septal cholinergic neurons which are preferentially vulnerable in AD pathology [190]. There is also evidence that memantine can reverse Aβ-induced LTP deficits, decrease Aβ secretion, and prevent Aβ-induced disruption of axonal transport, DNA fragmentation and microtubule deregulation [154, 191-194; Table 2]. Consistent with these data, several studies have demonstrated memantine can attenuate cognitive deficits, reduce Aβ-plaque burden, decrease levels of phosphorylated tau and simultaneously increase synaptic density in mutant APP transgenic mice overexpressing Aβ peptides [195-199]. Collectively, these results indicate that memantine, unlike many other NMDA receptor antagonists, is able to attenuate AD-related pathology as well as cognitive deficits without any major adverse effects, thus validating its therapeutic potential in the treatment of AD.

Table 2. Protective effects of memantine on AD-related behaviour and pathology

Beneficial effects of memantine	References
Neuroprotective effects against Aβ-induced toxicity	[146-148, 150,190]
Reversal of Aβ-induced long-term potentiation (LTP) deficits	[104, 191,194]
Beneficial effects of memantine	**References**
Attenuating cognitive deficits in mutant APP transgenic mice	[196-199]
Attenuating Aβ production/deposits and increasing synaptic density	[192, 196, 199]
Preventing Aβ-induced disruption of axonal transport/DNA fragmentation and microtubule degeneration	[154, 193]

The pharmacokinetic and neuroprotective profiles of memantine, together with some positive data from human clinical trials, convinced the European Union in 2002 and the Federal Drug Administration in the USA in 2003 to approve memantine for the treatment of moderate-to-severe AD patients [182, 184].

Over the years a number of additional trials and meta-analyses have shown beneficial effects of the drug on the global status, cognition and function of patients with moderate to severe AD without any major adverse effects [183, 200-202]. It has also been shown to improve behavioral disturbances such as agitation and aggression. Its mechanism of action as an NMDA receptor antagonist differs from the cholinesterase inhibitors approved for the treatment of AD patients (i.e., donepezil, galantamine and revastigmine), thus leading to the development of combinational therapeutic strategies for AD pathology [203]. Combination treatment strategies have proven to be well tolerated in randomized, controlled studies of patients with AD.

In some clinical trials with moderate to severe AD, treatment with memantine and donepezil resulted in significant benefits across cognitive, functional and behavioural domains over donepezil monotherapy [204-207]. On the other hand, when memantine/ cholinesterase inhibitors were used in patients with mild to moderate AD, the beneficial effects were not found to be statistically superior over the cholinesterase monotherapy [205]. A post-hoc meta-analysis, however, revealed significant benefits of the combination treatment compared to monotherapy for patients with moderate to severe AD but not when mild AD patients were included [208]. Overall the clinical data on efficacy of memantine, either as monotherapy or in conjunction with cholinesterase inhibitors, has been found to be somewhat less compelling for mild to moderate AD than those observed with moderate to severe AD [183, 203, 209, 210]. The side-effect profile of memantine is also very different from that of the cholinesterase inhibitors, thus allowing its use in AD patients who cannot tolerate cholinesterase inhibitors because of their gastrointestinal or cardiac adverse effects. Unfortunately, to date, clinical studies have examined the benefits of memantine with only limited doses of the drug. Given its mechanism of action as an NMDA receptor antagonist mediating its effect *via* the GluN2B subunit, it is of interest to determine whether higher doses of memantine with acceptable tolerability or the use of a more selective GluN2B antagonists might have more significant disease-modifying effects such as protection of neurons against excitotoxicity [38, 183, 195]. Some studies have suggested that analogues of memantine such as nitromemantines or memantine-sulfur containing antioxidant conjugates may be more neuroprotective than memantine in both *in vitro* and *in vivo* models [130, 211,

212]. However, more research is needed to define the efficacy of memantine analogues and to characterize other antagonists of GluN2B-containing NMDA receptors for use in AD patients.

Acknowledgment

The authors are grateful for funding from CIHR and from the estate of Ms. Irene Godberson to their research programs.

Abbreviations

ADAM	a disintegrin and metalloproteinase domain-containing peptidase;
Aβ	beta amyloid;
AD	Alzheimer's disease;
AICD	APP intracellular domain;
AMPA	α-amino-3-hydroxy-5-methyl-4-isoxazolepropionic acid;
APP	amyloid precursor protein;
CaMK	calmodulin-dependent protein kinase;
CDK	cyclin-dependent kinases;
CTF	C-terminal fragment;
CREB	cyclic AMP response element binding protein;
EAAT	excitatory amino acid transporter;
ERK	extracellular signal-related kinase;
GSK-3β	glycogen synthase kinase-3β;
iGluR	ionotropic glutamate receptor;
KA	kainate;
LTD	long term depression;
LTP	long term potentiation;
MCI	mild cognitive impairment;
mGluR	metabotropic glutamate receptor;
NFTs	neurofibrillary tangles;
NMDA	N-methyl-D-aspartate;
PSD	post-synaptic density;
$STEP_{61}$	striatal-enriched tyrosine phosphatase;
VGLUT	vesicular glutamate transporter

References

[1] Hugo, J. & Ganguli, M. (2014). Dementia and cognitive impairment: epidemiology, diagnosis and treatment. *Clin. Geriatr. Med.*, *30*, 421-442.

[2] Whitehouse, P. J. (1997). Genesis of Alzheimer's disease. *Neurology*, *48*, (5 Suppl 7), S2-S7.

[3] Selkoe, D. J. (2011). Alzheimer's disease. *Cold Spring Harbor PerspectivesBiol.*, *3*.

[4] Nelson, P. T., Braak, H. & Markesbery, W. R. (2009). Neuropathology and cognitive impairment in Alzheimer disease: a complex but coherent relationship. *J. Neuropathol. Exp. Neurol.*, *68*,1-14.

[5] Karch, C. M., Cruchaga, C. & Goate, A. M. (2014). Alzheimer's disease genetics: from the bench to the clinic. *Neuron*, *83*, 11-26.

[6] Hardy, J., Bogdanovic, N., Winblad, B., Portelius, E., Andreasen, N., Cedazo-Minguez, A. & Zetterberg, H. (2014). Pathways to Alzheimer's disease. *J. Int. Med.*, *275*, 296-303.

[7] St George-Hyslop, P. H. & Petit, A. (2005). Molecular biology and genetics of Alzheimer's disease. *C. R. Biol.*, *328*,119-130.

[8] Tanzi, R. E. & Bertram, L. (2005). Twenty years of the Alzheimer's disease amyloid hypothesis: a genetic perspective. *Cell*, *120*, 545-555.

[9] Coon, K. D., Myers, A. J., Craig, D. W., Webster, J. A., Pearson, J. V., Lince, D. H. Zismann, V. L., Beach, T. G., Leung, D., Bryden, L., Halperin, R. F., Marlowe, L., Kaleem, M., Walker, D. G., Ravid, R., Heward, C. B., Rogers, J., Papassotiropoulos, A., Reiman, E. M., Hardy, J. & Stephan, D. A. (2007). A high-density whole-genome association study reveals that APOE is the major susceptibility gene for sporadic late-onset Alzheimer's disease. *J. Clin. Psychiatry*, *68*, 613-618.

[10] Poirier, J. (2003). Apolipoprotein E and cholesterol metabolism in the pathogenesis and treatment of Alzheimer's disease. *Trends Mol. Med.*, *9*, 94-101.

[11] Harold, D., Abraham, R., Hollingworth, P., Sims, R., Gerrish, A., Hamshere, M. L., et al. (2009). Genome-wide association study identifies variants at CLU and PICALM associated with Alzheimer's disease. *Nat. Genet.*, *41*, 1088-1093.

[12] Lambert, J. C., Heath, S., Even, G., Campion, D., Sleegers, K., Hiltunen, M., et al. (2009). Genome-wide association study identifies variants at CLU and CR1 associated with Alzheimer's disease. *Nat. Genet.*, *41*, 1094-1099.

[13] Seshadri, S., Fitzpatrick, A. L., Ikram, M. A., DeStefano, A. L., Gudnason, V., Boada, M., et al. (2010). Genome-wide analysis of genetic loci associated with Alzheimer disease. *JAMA*, *303*, 1832-1840.

[14] Rogaeva, E., Meng, Y., Lee, J. H., Gu, Y., Kawarai, T., Zou, F., et al. (2007). The neuronal sortilin-related receptor SORL1 is genetically associated with Alzheimer disease. *Nat. Genet.*, *39*, 168-177.

[15] Hollingworth, P., Harold, D., Jones, L., Owen, M. J. & Williams, J. (2011). Alzheimer's disease genetics: current knowledge and future challenges. *Int. J. Geriatr. Psychiatry*, *26*, 793-802.

[16] Lambert, J. C. & Amouyel, P. (2011). Genetics of Alzheimer's disease: new evidences for an old hypothesis? *Curr. Opin. Genet. Dev.*, *21*, 295-301.

[17] Chen, J. H., Lin, K. P. & Chen, Y. C. (2009). Risk factors for dementia. *J. Formos. Med. Assoc.*, *108*, 754-764.

[18] Guskiewicz, K. M., Marshall, S. W., Bailes, J., McCrea, M., Cantu, R. C., Randolph, C. & Jordan, B. D. (2005). Association between recurrent concussion and late-life cognitive impairment in retired professional football players. *Neurosurgery*, *57*, 719-726.

[19] Querfurth, H. W. & LaFerla, F. M. (2010). Alzheimer's disease. *N. Engl. J. Med.*, *362*, 329-344.

[20] Brion, J. P., Anderton, B. H., Authelet, M., Dayanandan, R., Leroy, K., Lovestone, S., Octave, J. N., Pradier, L., Touchet, N. & Tremp, G. (2001). Neurofibrillary tangles and tau phosphorylation. *Biochem. Soc. Symp.*, *67*, 81-88.
[21] Wang, J. Z., Grundke-Iqbal, I. & Iqbal, K. (2007). Kinases and phosphatases and tau sites involved in Alzheimer neurofibrillary degeneration. *Eur. J. Neurosci.*, *25*, 59-68.
[22] Ballatore, C., Lee, V. M. & Trojanowski, J. Q. (2007). Tau-mediated neurodegeneration in Alzheimer's disease and related disorders. *Nat. Rev. Neurosci.*, *8*, 663-672.
[23] Mi, K. & Johnson, G. V. (2006). The role of tau phosphorylation in the pathogenesis of Alzheimer's disease. *Curr. Alz. Res.*, *3*, 449-463.
[24] Thal, D. R., Attems, J. & Ewers, M. (2014). Spreading of amyloid, tau and microvascular pathology in Alzheimer's disease: findings from neuropathological and neuroimaging studies. *Curr. Alz. Res.*, 42 (suppl 4), S421-429.
[25] Braak, H. & Braak, E. (1998). Evolution of neuronal changes in the course of Alzheimer's disease. *J. Neural. Transm. Suppl.*, *53*, 127-140.
[26] Schellenberg, G. D. & Montine, T. J. (2012). The genetics and neuropathology of Alzheimer's disease. *Acta Neuropathol.*, *124*, 305-323.
[27] Clippingdale, A. B., Wade, J. D. & Barrow, C. J. (2001). The amyloid-beta peptide and its role in Alzheimer's disease. *J. Pep. Sci.*, *7*, 227-249.
[28] Dickson, D. W. (1997). The pathogenesis of senile plaques. *J. Neuropathol. Exp. Neurol.*, *56*, 321-339.
[29] Naslund, J., Haroutunian, V., Mohs, R., Davis, K. L., Davies, P., Greengard, P. & Buxbaum, J. D. (2000). Correlation between elevated levels of amyloid beta-peptide in the brain and cognitive decline. *JAMA*, *283*, 1571-1577.
[30] Hardy, J., Bogdanovic, N., Winblad, B., Portelius, E., Andreasen, N., Cedazo-Minguez, A. & Zetterberg, H. (2014). Pathways to Alzheimer's disease. *J. Intern. Med.*, *275*, 296-303.
[31] Mawuenyega, K. G., Sigurdson, W., Ovod, V., Munsell, L., Kasten, T., Morris, J.C. Yarasheski, K. E. & Bateman, R. J. (2010). Decreased clearance of CNS beta-amyloid in Alzheimer's disease. *Science*, *330*, 1774.
[32] DeKosky, S. T., Scheff, S. W. & Styren, S. D. (1996). Structural correlates of cognition in dementia: quantification and assessment of synapse change. *Neurodegeneration*, *5*, 417-421.
[33] Kar, S., Slowikowski, S. P., Westaway, D. & Mount, H. T. (2004). Interactions between beta-amyloid and central cholinergic neurons: implications for Alzheimer's disease. *J. Psychiatry Neurosci.*, *29*, 427-441.
[34] Butterfield, D. A. 7 Pocernich, C. B. (2003). The glutamatergic system and Alzheimer's disease: therapeutic implications. *CNS Drugs*, *17*, 641-652.
[35] Francis, P. T. (2003). Glutamatergic systems in Alzheimer's disease. *Int. J. Geriatr. Psychiatry.*, *18*, S15-S21.
[36] Bell, K. F., Bennett, D. A. & Cuello, A. C. (2007). Paradoxical upregulation of glutamatergic presynaptic boutons during mild cognitive impairment. *J. Neurosci.*, *27*, 10810-10817.
[37] Bell, K. F. & Cuello, A. C. (2006). Altered synaptic function in Alzheimer's disease. *Eur. J. Pharmacol.*, *545*, 11-21.
[38] Dansyz, W. & Parsons, C. G. (2012). Alzheimer's disease, NMDA receptors and memantine – searching for the connections. *Brit. J. Pharmacol.*, *167*, 324-352.

[39] Esposito, Z., Belli, L., Toniolo, S., Sancesario, G., Bianconi, C. & Martorana, A. (2013). Amyloid β, glutamate, excitotoxicity in Alzheimer's disease: are we on the right track? *CNS Neurosci. Ther.*, *19*, 549-555.

[40] Revett, T. J., Baker, G. B., Jhamandas, J. & Kar, S. (2013). Glutamate system, amyloid ß peptides and tau protein: functional interrelationships and relevance to Alzheimer disease pathology. *J. Psychiatry Neurosci.*, *38*, 6-23.

[41] Fremeau, R. T., Jr. Troyer, M. D., Pahner, I., Nygaard, G. O., Tran, C. H., Reimer, R. J., Bellocchio, E. E., Fortin, D., Storm-Mathisen, J. & Edwards, R. H. (2001). The expression of vesicular glutamate transporters defines two classes of excitatory synapse. *Neuron*, *31*, 247-260.

[42] Kew, J. N. & Kemp, J. A. (2005) Ionotropic and metabotropic glutamate receptor structure and pharmacology. *Psychopharmacology (Berl)*, *179*, 4-29.

[43] Ozawa, S., Kamiya, H. & Tsuzuki, K. (1998). Glutamate receptors in the mammalian central nervous system. *Prog. Neurobiol.*, *54*, 581-618.

[44] Danbolt, N. C. (2001). Glutamate uptake. *Prog. Neurobiol.*, *65*, 1-105.

[45] Paoletti, P., Bellone, C. & Zhou, Q. (2013). NMDA receptor subunit diversity: impact on receptor properties, synaptic plasticity and disease. *Nat. Rev. Neurosci.*, *14*, 383-400.

[46] Anderson, C. M. & Swanson, R. A. (2000). Astrocyte glutamate transport: review of properties, regulation, and physiological functions. *Glia*, *32*, 1-14.

[47] Storck, T., Schulte, S., Hofmann, K. & Stoffel, W. (1992). Structure, expression, and functional analysis of a Na^+-dependent glutamate/aspartate transporter from rat brain. *Proc. Natl. Acad. Sci. USA*, *89*, 10955-10959.

[48] Pines, G., Danbolt, N. C., Bjoras, M., Zhang, Y., Bendahan, A., Eide, L., Koepsell, H., Storm-Mathisen, J., Seeberg, E. & Kanner, B. I. (1992). Cloning and expression of a rat brain L-glutamate transporter. *Nature*, *360*, 464-467.

[49] Shashidharan, P., Huntley, G. W., Meyer, T., Morrison, J. H. & Plaitakis, A. (1994). Neuron-specific human glutamate transporter: molecular cloning, characterization and expression in human brain. *Brain Res.*, *662*, 245-250.

[50] Zhou, Q. & Sheng, M. (2013). NMDA receptors in nervous system diseases. *Neuropharmacol.*, *74*, 69-75.

[51] Hardingham, G. E. & Bading, H. (2010). Synaptic versus extrasynaptic NMDA receptor signaling: implications for neurodegenerative disorders. *Nat. Rev. Neurosci.*, *11*, 682-696.

[52] Ferraguti, F. & Shigemoto, R. (2006). Metabotropic glutamate receptors. *Cell Tissue Res.*, *326*, 483-504.

[53] Gerber, U., Gee, C. E. & Benquet, P. (2007). Metabotropic glutamate receptors:intracellular signaling pathways. *Curr. Opin. Pharmacol.*, *7*, 56-61.

[54] Schools, G. P. & Kimelberg, H. K. (1999). mGluR3 and mGluR5 are the predominant metabotropic glutatmate receptor mRNAs expressed in hippocampal astrocytes acutely isolated from young rats. *J. Neurosci. Res.*, *58*, 533-543.

[55] Kowall, N. W. & Beal, M. F. (1991). Glutamate-, glutaminase-, and taurine-immunoreactive neurons develop neurofibrillary tangles in Alzheimer's disease. *Ann. Neurol.*, *29*, 162-167.

[56] Bussiere, T., Giannakopoulos, P., Bouras, C., Perl, D. P., Morrison, J. H. & Hof, P. R. (2003). Progressive degeneration of nonphosphorylated neurofilament protein-enriched

pyramidal neurons predicts cognitive impairment in Alzheimer's disease: stereologic analysis of prefrontal cortex area 9. *J. Comp. Neurol.*, *463*, 281-302.
[57] Kashani, A., Lepicard, E., Poirel, O., Videau, C., David, J. P., Fallet-Bianco, C., Simon, A., Delacourte, A., Giros, B., Epelbaum, J., Betancur, C. & El Mestikawy, S. (2008). Loss of VGLUT1 and VGLUT2 in the prefrontal cortex is correlated with cognitive decline in Alzheimer disease. *Neurobiol. Aging*, *29*, 1619-1630.
[58] Kirvell, S. L., Esiri, M. & Francis, P. T. (2006). Down-regulation of vesicular glutamate transporters precedes cell loss and pathology in Alzheimer's disease. *J. Neurochem.*, *98*, 939-950.
[59] Sokolow, S., Luu, S. H., Nandy, K., Miller, C. A., Vinters, H. V., Poon, W. W. & Gylys, K. H. (2012). Preferential accumulation of amyloid □□in presynaptic glutamatergic terminals (VGLUT1 and VGLUT2) in Alzheimer's disease cortex. *Neurobiol. Dis.*, *45*, 381-387.
[60] Jacob, C. P., Koutsilieri, E., Bartl, J., Neuen-Jacob, E., Arzberger, T., Zander, N., Ravid, R., Roggendorf, W., Riederer, P. & Grünblatt, E. (2007). Alterations in expression of glutamatergic transporters and receptors in sporadic Alzheimer's disease. *J. Alz. Dis.*, *11*, 97-116.
[61] Scott, H. A., Gebhardt, F. M., Mitrovic, A. D., Vandenberg, R. J. & Dodd, P. R. (2011). Glutamate transporter variants reduce glutamate uptake in Alzheimer's disease. *Neurobiol. Aging.*, *32*, 553.e1-11.
[62] Masliah, E., Alford, M., DeTeresa, R., Mallory, M. & Hansen, L. (1996). Deficient glutamate transport is associated with neurodegeneration in Alzheimer's disease. *Ann. Neurol.*, *40*, 759-766.
[63] Li, S., Mallory, M., Alford, M., Tanaka, S. & Masliah, E. (1997). Glutamate transporter alterations in Alzheimer's disease are possibly associated with abnormal APP expression. *J. Neuropathol. Exp. Neurol.*, *56*, 901-911.
[64] Robinson, S. R. (2001). Changes in the cellular distribution of glutamine synthetase in Alzheimer's disease. *J. Neurosci. Res.*, *66*, 972-980.
[65] Ong, W. Y., Tanaka, K., Dawe, G. S., Ittner, L. M., Farooqui, A. A. (2013). Slow excitotoxicity in Alzheimer's disease. *J, Alzheimers Dis.*, *35*, 643-668.
[66] Arundine, M. & Tymianski, M. (2003). Molecular mechanisms of calcium-dependent neurodegeneration in excitotoxicity. *Cell Calcium*, *34*, 325-337.
[67] Dong, X. X., Wang, Y. & Qin, Z. H. (2009). Molecular mechanisms of excitotoxicity and their relevance to pathogenesis of neurodegenerative diseases. *Acta Pharmacol. Sin.*, *30*, 379-387.
[68] Vosler, P. S., Brennan, C. S. & Chen, J. (2008). Calpain-mediated signaling mechanisms in neuronal injury and neurodegeneration. *Mol. Neurobiol.*, *38*, 78-100.
[69] Ulas, J., Weihmuller, F. B., Brunner, L. C., Joyce, J. N., Marshall, J. F. & Cotman, C. W. (1994). Selective increase of NMDA-sensitive glutamate binding in the striatum of Parkinson's disease, Alzheimer's disease, and mixed Parkinson's disease/Alzheimer's disease patients: an autoradiographic study. *J. Neurosci.*, *14*, 6317-6324.
[70] Ikonomovic, M. D., Mizukami, K., Warde, D., Sheffield, R., Hamilton, R., Wenthold, R. J. & Armstrong, D. M. (1999). Distribution of glutamate receptor subunit NMDAR1 in the hippocampus of normal elderly and patients with Alzheimer's disease. *Exp. Neurol.*, *160*, 194-204.

[71] Panegyres, P. K., Zafiris-Toufexis, K. & Kakulas, B. A. (2002). The mRNA of the NR1 subtype of glutamate receptor in Alzheimer's disease. *J. Neural. Transm.*, *109*, 77-89.

[72] Sze, C., Bi, H., Kleinschmidt-DeMasters, B. K., Filley, C. M. 7 Martin, L. J. (2001). N-Methyl-D-aspartate receptor subunit proteins and their phosphorylation status are altered selectively in Alzheimer's disease. *J. Neurol. Sci.*, *182*, 151-159.

[73] Hynd, M. R., Scott, H. L. & Dodd, P. R. (2004). Selective loss of NMDA receptor NR1 subunit isoforms in Alzheimer's disease. *J. Neurochem.*, *89*, 240-247.

[74] Hynd, M. R., Scott, H. L. & Dodd, P. R. (2004). Differential expression of N-methyl D-aspartate receptor NR2 isoforms in Alzheimer's disease. *J. Neurochem.*, *90*, 913-919.

[75] Williams, C., Mehrian Shai, R., Wu, Y., Hsu, Y. H., Sitzer, T., Spann, B., McCleary, C., Mo, Y. & Miller, C. A. (2009). Transcriptome analysis of synaptoneurosomes identifies neuroplasticity genes overexpressed in incipient Alzheimer's disease. *PLoS One*, *4*, e4936.

[76] Pellegrini-Giampietro, D. E., Bennett, M. V. & Zukin, R. S. (1994). AMPA/kainate receptor gene expression in normal and Alzheimer's disease hippocampus. *Neuroscience*, *61*, 41-49.

[77] Yasuda, R. P., Ikonomovic, M. D., Sheffield, R., Rubin, R. T., Wolfe, B. B. & Armstrong, D. M. (1995). Reduction of AMPA-selective glutamate receptor subunits in the entorhinal cortex of patients with Alzheimer's disease pathology: a biochemical study. *Brain Res.*, *678*, 161-167.

[78] Dewar, D., Chalmers, D. T., Graham, D. I. & McCulloch, J. (1991). Glutamate metabotropic and AMPA binding sites are reduced in Alzheimer's disease: an autoradiographic study of the hippocampus. *Brain Res.*, *553*, 58-64.

[79] Aronica, E., Dickson, D. W., Kress, Y., Morrison, J. H. & Zukin, R. S. (1998). Non-plaque dystrophic dendrites in Alzheimer hippocampus: a new pathological structure revealed by glutamate receptor immunocytochemistry. *Neuroscience*, *82*, 979-991.

[80] Albasanz, J. L., Dalfo, E., Ferrer, I. & Martin, M. (2005). Impaired metabotropic glutamate receptor/ phospholipase C signaling pathway in the cerebral cortex in Alzheimer's disease and dementia with Lewy bodies correlates with stage of Alzheimer's-disease-related changes. *Neurobiol. Dis.*, *20*, 685-693.

[81] Lee, H. G., Ogawa, O., Zhu, X., O'Neill, M. J., Petersen, R. B., Castellani, R. J., Ghanbari, H., Perry, G. & Smith, M. A. (2004). Aberrant expression of metabotropic glutamate receptor 2 in the vulnerable neurons of Alzheimer's disease. *Acta Neuropathol.*, *107*, 365-371.

[82] Allinson, T. M., Parkin, E. T., Turner, A. J. & Hooper, N. M. (2003). ADAMs family members as amyloid precursor protein alpha-secretases. *J. Neurosci. Res.*, *74*, 342-352.

[83] Chow, V. W., Mattson, M. P., Wong, P. C. & Gleichmann, M. (2010). An overview of APP processing enzymes and products. *Neuromolecular Med.*, *12*, 1-12.

[84] Gilbert, B. J. (2013). The role of amyloid β in the pathogenesis of Alzheimer's disease. *J. Clin. Pathol.*, *66*, 362-366.

[85] Bordji, K., Becerril-Ortega, J. & Buisson, A. (2011). Synapses, NMDA receptor activity and neuronal Aβ production in Alzheimer's disease. *Rev. Neurosci.*, *22*, 285-294.

[86] Roberson, M. R. & Harrell, L. E. (1997). Cholinergic activity and amyloid precursor protein metabolism. *Brain Res. Rev.*, *25*, 50-69.

[87] Hoey, S. E., Williams, R. & Perkinton, M. S. (2009). Synaptic NMDA receptor activation stimulates α–secretase amyloid precursor protein processing and inhibits amyloid–β production. *J. Neurosci.*, *29*, 4442-4460.

[88] Lee, R. K., Jimenez, J., Cox, A. J. & Wurtman, R. J. (1996). Metabotropic glutamate receptors regulate APP processing in hippocampal neurons and cortical astrocytes derived from fetal rats. *Ann. N. Y. Acad. Sci.*, *777*, 338-343.

[89] Bordji, K., Becerril-Ortega, J., Nicole, O. & Buisson, A. (2010). Activation of extrasynaptic, but not synaptic, NMDA receptors modifies amyloid precursor protein expression pattern and increases amyloid-β production. *J. Neurosci.*, *30*, 15927-15942.

[90] Lesne, S., Ali, C., Gabriel, C., Croci, N., MacKenzie, E. T., Glabe, C. G., Plotkine, M., Marchand-Verrecchia, C., Vivien, D. & Buisson, A. (2005). NMDA receptor activation inhibits α-secretase and promotes neuronal amyloid-β production. *J. Neurosci.*, *25*, 9367-9377.

[91] Verges, D. K., Restivo, J. L., Goebel, W. D., Holtzman, D. M. & Cirrito, J. R. (2011). Opposing synaptic regulation of amyloid-β metabolism by NMDA receptors *in vivo*. *J. Neurosci. 31*, 11328-11337.

[92] Hoey, S. E., Buonocore, F., Cox, C. J., Hammond, V. J., Perkinton, M. S. & Williams, R. J. (2013). AMPA receptor activation promotes non-amyloidogenic amyloid precursor protein processing and suppresses neuronal amyloid-β production. *PLoS One*, 8, e78155.

[93] Kim, S. H., Fraser, P. E., Westaway, D., St George-Hyslop, P. H., Ehrlich, M. E. & Gandy, S. (2010). Group II metabotropic glutamate receptor stimulation triggers production and release of Alzheimer's amyloidβ42 from isolated intact nerve terminals. *J. Neurosci.*, *30*, 3870-3875.

[94] Tang, B. L. (2009). Neuronal protein trafficking associated with Alzheimer disease: from APP and BACE1 to glutamate receptors. *Cell Adh. Migr.*, *3*, 118-128.

[95] Kullmann, D. M. & Lamsa, K. P. (2007). Long-term synaptic plasticity in hippocampal inter-neurons. *Nat. Rev. Neurosci.*, *8*, 687-699.

[96] Morris, R. G. (2003). Long-term potentiation and memory. *Philos. Trans. R. Soc. Lond. B. Biol. Sci.*, *358*, 643-647.

[97] Shankar, G. M., Li, S., Mehta, T. H., Garcia-Munoz, A., Shepardson, N. E., Smith, I., Brett, F. M., Farrell, M. A., Rowan, M. J., Lemere, C. A., Regan, C. M., Walsh, D. M., Sabatini, B. L. & Selkoe, D. J. (2008). Amyloid-β protein dimers isolated directly from Alzheimer's brains impair synaptic plasticity and memory. *Nat. Med.*, *14*, 837-842.

[98] Cullen, W. K., Suh, Y. H., Anwyl, R. & Rowan, M. J. (1997). Block of LTP in rat hippocampus *in vivo* by β-amyloid precursor protein fragments. *Neuroreport*, *8*, 3213-3217.

[99] Parihar, M. S. & Brewer, G. J. (2010). Amyloid-β as a modulator of synaptic plasticity. *J. Alz. Dis.*, *22*, 741-763.

[100] Klyubin, I., Betts, V., Welzel, A. T., Blennow, K., Zetterberg, H., Wallin, A., Lemere, C. A., Cullen, W. K., Peng, Y., Wisniewski, T., Selkoe, D. J., Anwyl, R., Walsh, D. M. & Rowan, M. J. (2008). Amyloid β protein dimer-containing human CSF disrupts synaptic plasticity: prevention by systemic passive immunization. *J. Neurosci.*, *28*, 4231-4237.

[101] Shankar, G. M., Bloodgood, B. L., Townsend, M., Walsh, D. M., Selkoe, D. J. & Sabatini, B. L. (2007). Natural oligomers of the Alzheimer amyloid-β protein induce reversible synapse loss by modulating an NMDA-type glutamate receptor-dependent signaling pathway. *J. Neurosci.*, *27*, 2866-2875.

[102] Li, S., Jin, M., Koeglsperger, T., Shepardson, N. E., Shankar, G. M. & Selkoe, D. J. (2011). Soluble Aβ oligomers inhibit long-term potentiation through a mechanism involving excessive activation of extrasynaptic NR2B-containing NMDA receptors. *J. Neurosci.*, *31*, 6627-6638.

[103] Selkoe, D. J. (2008). Soluble oligomers of the amyloid β-protein impair synaptic plasticity and behavior. *Behav. Brain Res.*, *192*, 106-113.

[104] Hu, N. W., Klyubin, I., Anwyl, R. & Rowan, M. J. (2009). GluN2B subunit-containing NMDA receptor antagonists prevent Abeta-mediated synaptic plasticity disruption *in vivo*. *Proc. Natl. Acad. Sci. USA*, *106*, 20504-20509.

[105] Li, S., Hong, S., Shepardson, N. E., Walsh, D. M., Shankar, G. M. & Selkoe, D. (2009). Soluble oligomers of amyloid Beta protein facilitate hippocampal long-term depression by disrupting neuronal glutamate uptake. *Neuron*, *62*, 788-801.

[106] Cullen, W. K., Wu, J., Anwyl, R. & Rowan, M. J. (1996). beta-Amyloid produces a delayed NMDA receptor-dependent reduction in synaptic transmission in rat hippocampus. *Neuroreport*, *8*, 87-92.

[107] Kessels, H. W., Nabavi, S. & Malinow, R. (2013). Metabotropic NMDA receptor function is required for β-amyloid-induced synaptic depression. *Proc. Natl. Acad. Sci., USA*, *110*, 4033-4308.

[108] Tamburri, A., Dudilot, A., Licea, S., Bourgeois, C. & Boehm, J. (2013). NMDA-receptor activation but not ion flux is required for amyloid-beta induced synaptic depression. *PLoS One*, *8*, e65350.

[109] Wei, W., Nguyen, L. N., Kessels. H. W., Hagiwara, H., Sisodia, S. & Malinow, R. (2010). Amyloid β from axons and dendrites reduces local spine number and plasticity. *Nat. Neurosci.*, *13*, 190-196.

[110] Knafo, S., Alonso-Nanclares, L., Gonzalez-Soriano, J., Merino-Serrais, P., Fernaud-Espinosa, I., Ferrer, I. & DeFelipe, J. (2009). Widespread changes in dendritic spines in a model of Alzheimer's disease. *Cereb. Cortex*, *19*, 586-592.

[111] Cirrito, J. R., Yamada, K. A., Finn, M. B., Sloviter, R. S., Bales, K. R., May, P. C., Schoepp, D. D., Paul, S. M., Mennerick, S. & Holtzman, D. M. (2005). Synaptic activity regulates interstitial fluid amyloid-β levels *in vivo*. *Neuron*, *48*, 913-922.

[112] Ting, J. T., Kelley, B. G., Lambert, T. J., Cook, D. G. & Sullivan, J. M. (2007). Amyloid precursor protein overexpression depresses excitatory transmission through both presynaptic and postsynaptic mechanisms. *Proc. Natl. Acad. Sci. USA*, *104*, 353-358.

[113] Snyder, E. M., Nong, Y., Almeida, C. G., Paul, S., Moran, T., Choi, E. Y., Nairn, A. C., Salter, M. W., Lombroso, P. J., Gouras, G. K. & Greengard, P. (2005). Regulation of NMDA receptor trafficking by amyloid-β. *Nat. Neurosci.*, *8*, 1051-1058.

[114] Braithwaite, S. P., Adkisson, M., Leung, J., Nava, A., Msterson, B., Urfer, R., Oksenberg, D. & Nikolich, K. (2006). Regulation of NMDA receptor trafficking and function by striatal-enriched tyrosine phosphatase (STEP). *Eur. J. Neurosci.*, *23*, 2847-2856.

[115] Kurup, P., Zhang, Y., Venkitaramani, D. V., Xu, J., Lombroso, P. J. (2010). The role of STEP in Alzheimer's disease. *Channels (Austin)*, *4*, 347-350.

[116] Zhang, Y., Venkitaramani, D. V., Gladding, C. M., Zhang, Y., Kurup, P., Molnar, E., Collingridge, G. L. & Lombroso, P. J. (2008). The tyrosine phosphatase STEP mediates AMPA receptor endocytosis after metabotropic glutamate receptor stimulation. *J. Neurosci.*, *28*, 10561-10566.

[117] Gong, Y. & Lippa, C. F. (2010). Review: disruption of the postsynaptic density in Alzheimer's disease and other neurodegenerative dementias. *Am. J. Alzheimers Dis. Other Demen.*, *25*, 547-555.

[118] Roselli, F., Hutzler, P., Wegerich, Y., Livrea, P. & Almeida, O. F. (2009). Disassembly of shank and homer synaptic clusters is driven by soluble β-amyloid(1-40) through divergent NMDAR-dependent signaling pathways. *PLoS One*, *4*, e6011.

[119] Ehrlich, I., Klein, M., Rumpel, S. & Malinow, R. (2007). PSD-95 is required for activity-driven synapse stabilization. *Proc. Natl. Acad. Sci. USA*, *104*, 4176-4181.

[120] Kammermeier, P. J. (2006). Surface clustering of metabotropic glutamate receptor 1 induced by long Homer proteins. *BMC Neurosci.*, *7*, 1.

[121] Farr, S. A., Banks, W. A., Uezu, K., Sano, A., Gaskin, F. S. & Morley, J. E. (2003). Antibody to β-amyloid protein increases acetylcholine in the hippocampus of 12 month SAMP8 male mice. *Life Sci.*, *73*, 555-562.

[122] Kar, S., Seto, D., Gaudreau, P. & Quirion, R. (1996). β-amyloid-related peptides inhibit potassium-evoked acetylcholine release from rat hippocampal slices. *J. Neurosci.*, *16*, 1034-1040.

[123] Bobich, J. A., Zheng, Q. & Campbell, A. (2004). Incubation of nerve endings with a physiological concentration of Aβ1-42 activates CaV2.2(N-Type)-voltage operated calcium channels and acutely increases glutamate and noradrenaline release. *J. Alzheimers Dis.*, *6*, 243-255.

[124] Li, M. & Smith, C. P. (1996). β-amyloid1-40 inhibits electrically stimulated release of $[^3H]$norepinephrine and enhances the internal calcium response to low potassium in rat cortex: prevention with a free radical scavenger. *Brain Res. Bull.*, *39*, 299-303.

[125] Arias, C., Arrieta, I. & Tapia, R. (1995). β-Amyloid peptide fragment 25-35 potentiates the calcium-dependent release of excitatory amino acids from depolarized hippocampal slices. *J. Neurosci. Res.*, *41*, 561-566.

[126] Mura, E., Preda, S., Govoni, S., Lanni, C., Trabace, L., Grilli, M., Lagomarsino, F., Pittaluga, A. & Marchi, M. (2010). Specific neuromodulatory actions of amyloid-beta on dopamine release in rat nucleus accumbens and caudate putamen. *J. Alzheimers Dis.*, *19*, 1041-1053.

[127] Kabogo, D., Rauw, G., Amritraj, A., Baker, G. & Kar, S. (2010). β-amyloid-related peptides potentiate K^+-evoked glutamate release from adult rat hippocampal slices. *Neurobiol. Aging*, *31*, 1164-1172.

[128] Brito-Moreira, J., Paula-Lima, A. C., Bomfim, T. R., Oliveira, F. B., Sepúlveda, F. J., De Mello, F. G., Aguayo, L. G., Panizzutti, R. & Ferreira, S. T. (2011). Aβ oligomers induce glutamate release from hippocampal neurons. *Curr. Alzheimer Res.*, *8*, 552-562.

[129] Salamone, A., Mura, E., Zappettini, S., Grilli, M., Olivero, G., Preda, S., Govoni, S. & Marchi, M. (2014). Inhibitory effects of beta-amyloid on the nicotinic receptors which

stimulate glutamate release in rat hippocampus: the glial contribution. *Eur. J. Pharmacol.*, *723*, 314-321.

[130] Talantova, M., Sanz-Blasco, S., Zhang, X., Xia, P., Akhtar, M. W., Okamoto, S., Dziewczapolski, G., Nakamura, T., Cao, G., Pratt, A. E., Kang, Y. J., Tu, S., Molokanova, E., McKercher, S. R., Hires, S. A., Sason, H., Stouffer, D. G., Buczynski, M. W., Solomon, J. P., Michael, S., Powers, E. T., Kelly, J. W., Roberts, A., Tong, G., Fang-Newmeyer, T., Parker, J., Holland, E. A., Zhang, D., Nakanishi, N., Chen, H. S., Wolosker, H., Wang, Y., Parsons, L. H., Ambasudhan, R., Masliah, E., Heinemann, S. F., Piña-Crespo, J. C. & Lipton, S. A. (2013). Aβ induces astrocytic glutamate release, extrasynaptic NMDA receptor activation, and synaptic loss. *Proc. Natl. Acad. Sci. USA*, *110*, E2518-E2527.

[131] Hascup, K. N. & Hascup, E. R. (2015). Altered neurotransmission prior to cognitive decline in AβPP/PS1 mice, a model of Alzheimer's disease. *J. Alzheimers Dis.*, *44*, 771-776.

[132] Abe, K. & Misawa, M. (2003). The extracellular signal-regulated kinase cascade suppresses amyloid β protein-induced promotion of glutamate clearance in cultured rat cortical astrocytes. *Brain Res.*, *979*, 179-187.

[133] Rodriguez-Kern, A., Gegelashvili, M., Schousboe, A., Zhang, J., Sung, L. & Gegetashvili. (2003). β-amyloid and brain-derived neurotrophic factor, BDNF, up-regulate the expression of glutamate transporter GLT-1/EAAT2 *via* different signaling pathways utilizing transcription factor NF-kappaB. *Neurochem. Int.*, *43*, 363-370.

[134] Ikegaya, Y., Matsuura, S., Ueno, S., Baba, A., Yamada, M. K., Nishiyama, N. & Matsuki, N. (2002). Beta-amyloid enhances glial glutamate uptake activity and attenuates synaptic efficacy. *J. Biol. Chem.*, *277*, 32180-32186.

[135] Matos, M., Augusto, E., Oliveira, C. R. & Agostinho, P. (2008). Amyloid-β peptide decreases glutamate uptake in cultured astrocytes: involvement of oxidative stress and mitogen-activated protein kinase cascades. *Neuroscience*, *156*, 898-910.

[136] Fernandez-Tome, P., Brera, B., Arevalo, M. A. & de Ceballos, M. L. (2004). β-amyloid25-35 inhibits glutamate uptake in cultured neurons and astrocytes: modulation of uptake as a survival mechanism. *Neurobiol. Dis.*, *15*, 580-589.

[137] Harris, M. E., Wang, Y., Pedigo, N.W. Jr, Hensley, K., Butterfield, D. A. & Carney, J. M. (1996). Amyloid β peptide (25-35) inhibits Na^+-dependent glutamate uptake in rat hippocampal astrocyte cultures. *J. Neurochem.*, *67*, 277-286.

[138] Lauderback, C. M., Hackett, J. M., Huang, F. F., Keller, J. N., Szweda, L. I., Markesbery, W. R. & Butterfield, D. A. (2001). The glial glutamate transporter, GLT-*1*, is oxidatively modified by 4-hydroxy-2-nonenal in the Alzheimer's disease brain: the role of Aβ1-*J. Neurochem.*, *78*, 413-416.

[139] Noda, M., Nakanishi, H. & Akaike, N. (1999). Glutamate release from microglia *via* glutamate transporter is enhanced by amyloid-β peptide. *Neuroscience*, *92*, 1465-1474.

[140] Kayed, R. & Lasagna-Reeves, C. A. (2013). Molecular mechanisms of amyloid oligomers toxicity. *J. Alzheimers Dis.*, 33 (suppl 1), S67-78.

[141] Yankner, B. A. (1996). Mechanisms of neuronal degeneration in Alzheimer's disease. *Neuron*, *16*, 921-932.

[142] Smith, W. W., Gorospe, M. & Kusiak, J. W. (2006). Signaling mechanisms underlying Aβ toxicity: potential therapeutic targets for Alzheimer's disease. *CNS Neurol. Disord. Drug Targets*, *5*, 355-361.

[143] Blurton-Jones, M. & Laferla, F. M. (2006). Pathways by which Aβ facilitates tau pathology. *Curr. Alzheimer Res.*, *3*, 437-448.

[144] Roth, K. A. (2001). Caspases, apoptosis, and Alzheimer disease: causation, correlation, and confusion. *J. Neuropathol. Exp. Neurol.*, *60*, 829-838.

[145] Mattson, M. P., Cheng, B., Davis, D., Bryant, K., Lieberburg, I. & Rydel, R. E. (1992). β-Amyloid peptides destabilize calcium homeostasis and render human cortical neurons vulnerable to excitotoxicity. *J. Neurosci.*, *12*, 376-389.

[146] Floden, A. M., Li, S. & Combs, C. K. (2005). β–amyloid-stimulated microglia induce neuron death *via* synergistic stimulation of tumor necrosis factor α and NMDA receptors. *J. Neurosci.*, *25*, 2566-2575.

[147] Song, M. S., Rauw, G., Baker, G. B. & Kar, S. (2008). Memantine protects rat cortical cultured neurons against β-amyloid-induced toxicity by attenuating tau phosphorylation. *Eur. J. Neurosci.*, *28*, 1989-2002.

[148] Tremblay, R., Chakravarthy, B., Hewitt, K., Tauskela, J., Morley, P., Atkinson, T. & Durkin, J. P. (2000). Transient NMDA receptor inactivation provides long-term protection to cultured cortical neurons from a variety of death signals. *J. Neurosci.*, *20*, 7183-7192.

[149] Harkany, T., Abraham, I., Timmerman, W., Laskay, G., Tóth, B., Sasvári, M., Kónya, C., Sebens, J. B., Korf, J., Nyakas, C., Zarándi, M., Soós, K., Penke, B. & Luiten, P. G. (2000). β-amyloid neurotoxicity is mediated by a glutamate-triggered excitotoxic cascade in rat nucleus basalis. *Eur. J. Neurosci.*, *12*, 2735-2745.

[150] Miguel-Hidalgo, J. J., Alvarez, X. A., Cacabelos, R. & Quack, G. (2002). Neuroprotection by memantine against neurodegeneration induced by β-amyloid(1-40). *Brain Res.*, *958*, 210-221.

[151] Guo, Q., Fu, W., Sopher, B. L., Miller, M. W., Ware, C. B., Martin, G. M. & Mattson, M. P. (1999). Increased vulnerability of hippocampal neurons to excitotoxic necrosis in presenilin-1 mutant knock-in mice. *Nat. Med.*, *5*, 101-106.

[152] Moechars, D., Lorent, K. & Van Leuven, F. (1999). Premature death in transgenic mice that overexpress a mutant amyloid precursor protein is preceded by severe neurodegeneration and apoptosis. *Neuroscience*, *91*, 819-830.

[153] Chin, J. H., Ma, L., MacTavish, D. & Jhamandas, J. H. (2007). Amyloid β protein modulates glutamate-mediated neurotransmission in the rat basal forebrain: involvement of presynaptic neuronal nicotinic acetylcholine and metabotropic glutamate receptors. *J. Neurosci.*, *27*, 9262-9926.

[154] Mota, S. I., Ferreira, I. L., Pereira, C., Oliveira, C. R. & Rego, A. C. (2012). Amyloid-β peptide 1-42 causes microtubule deregulation through N-methyl-D-aspartate receptors in mature hippocampal cultures. *Curr. Alzheimer Res.*, *9*, 844-856.

[155] Costa, R. O., Lacor, P. N., Ferreira, I. L., Resende, R., Auberson, Y. P., Klein, W. L., Oliveira, C. R., Rego, A. C. & Pereira, C. M. (2012). Endoplasmic reticulum stress occurs downstream of GluN2B subunit of N-methyl-D-aspartate receptor in mature hippocampal cultures treated with amyloid-β oligomers. *Aging Cell*, *11*, 823-833.

[156] Tackenberg, C., Grinschgl, S., Trutzel, A., Santuccione, A. C., Frey, M. C., Konietzko, U., Grimm, J., Brandt, R. & Nitsch, R. M. (2013). NMDA receptor subunit composition determines beta-amyloid-induced neurodegeneration and synaptic loss. *Cell Death Dis.*, *4*, e608.

[157] Liu, Y., Wong, T. P., Aarts, M., Rooyakkers, A., Liu, L., Lai, T. W., Wu, D. C., Lu, J., Tymianski, M., Craig, A. M. & Wang, Y. T. (2007). NMDA receptor subunits have differential roles in mediating excitotoxic neuronal death both *in vitro* and *in vivo*. *J. Neurosci.*, *27*, 2846-2857.

[158] Inoue, R., Hashimoto, K., Harai, T. & Mori, H. (2008). NMDA- and beta-amyloid1-42-induced neurotoxicity is attenuated in serine racemase knock-out mice. *J. Neurosci.*, *28*, 14486-14491.

[159] Hernandez, P., Lee, G., Sjoberg, M. & Maccioni, R. B. (2009). Tau phosphorylation by cdk5 and Fyn in response to amyloid peptide Aβ (25-35): involvement of lipid rafts. *J. Alzheimers Dis.*, *16*, 149-156.

[160] Ittner, L. M., Ke, Y. D., Delerue, F., Bi, M., Gladbach, A., van Eersel, J., Wölfing, H., Chieng, B. C., Christie, M. J., Napier, I. A., Eckert, A., Staufenbiel, M., Hardeman, E. & Götz, J. (2010). Dendritic function of tau mediates amyloid-β toxicity in Alzheimer's disease mouse models. *Cell*, *142*, 387-397.

[161] Liu, F., Grundke-Iqbal, I., Iqbal, K. & Gong, C. X. (2005). Contributions of protein phosphatases PP*1*, PP2A, PP2B and PP5 to the regulation of tau phosphorylation. *Eur. J. Neurosci.*, *22*, 1942-1950.

[162] Liu, F., Iqbal, K., Grundke-Iqbal, I., Rossie, S. & Gong, C. X. (2005). Dephosphorylation of tau by protein phosphatase 5: impairment in Alzheimer's disease. *J. Biol. Chem.*, *280*, 1790-1796.

[163] Gong, C. X., Shaikh, S., Wang, J. Z., Zaidi, T., Grundke-Iqbal, I. & Iqbal, K. (1995). Phosphatase activity toward abnormally phosphorylated tau: decrease in Alzheimer disease brain. *J. Neurochem.*, *65*, 732-738.

[164] Sontag, E., Luangpirom, A., Hladik, C., Mudrak, I., Ogris, E., Speciale, S. & White, C. L. (2004). Altered expression levels of the protein phosphatase 2A ABalphaC enzyme are associated with Alzheimer disease pathology. *J. Neuropathol. Exp. Neurol.*, *63*, 287-301.

[165] Vogelsberg-Ragaglia, V., Schuck, T., Trojanowski, J. Q. & Lee, V. M. (2001). PP2A mRNA expression is quantitatively decreased in Alzheimer's disease hippocampus. *Exp. Neurol.*, *168*, 402-412.

[166] Bennecib, M., Gong, C. X., Grundke-Iqbal, I. & Iqbal, K. (2001). Inhibition of PP-2A upregulates CaMKII in rat forebrain and induces hyperphosphorylation of tau at Ser 262/3*FEBS Lett.*, *490*, 15-22.

[167] Gong, C. X., Lidsky, T., Wegiel, J.,Zuck, L., Grundke-Iqbal, I. & Iqbal, K. (2000). Phosphorylation of microtubule-associated protein tau is regulated by protein phosphatase 2A in mammalian brain. Implications for neurofibrillary degeneration in Alzheimer's disease. *J. Biol. Chem.*, *275*, 5535-5544.

[168] Qian, W., Shi, J., Yin, X., Iqbal, K., Grundke-Iqbal, I., Gong, C. X. & Liu, F. (2010). PP2A regulates tau phosphorylation directly and also indirectly via activating GSK-3β. *J. Alzheimers Dis.*, *19*, 1221-1229.

[169] Pei, J. J., Gong, C. X., An, W. L., Winblad, B., Cowburn, R. F., Grundke-Iqbal, I. & Iqbal, K. (2003). Okadaic-acid-induced inhibition of protein phosphatase 2A produces activation of mitogen-activated protein kinases ERK1/2, MEK1/2, and p70 S6, similar to that in Alzheimer's disease. *Am. J. Pathol.*, *163*, 845-858.

[170] Cotman, C. W., Poon, W. W., Rissman, R. A. & Blurton-Jones, M. (2005). The role of caspase cleavage of tau in Alzheimer disease neuropathology. *J. Neuropathol. Exp. Neurol.*, *64*, 104-112.

[171] Gamblin, T. C., Chen, F., Zambrano, A., Abraha, A., Lagalwar, S., Guillozet, A. L., Lu, M., Fu, Y., Garcia-Sierra, F., LaPointe, N., Miller, R., Berry, R., Binder, L. & Cryns, V. (2003). Caspase cleavage of tau: linking amyloid and neurofibrillary tangles in Alzheimer's disease. *Proc. Natl. Acad. Sci. USA*, *100*, 10032-10037.

[172] Rissman, R. A., Poon, W. W., Blurton-Jones, M., Oddo, S., Torp, R., Vitek, M. P., LaFerla, F. M., Rohn, T. T. & Cotman, C. W. (2004). Caspase-cleavage of tau is an early event in Alzheimer disease tangle pathology. *J. Clin. Invest.*, *114*, 121-130.

[173] Liu, M. C., Kobeissy, F., Zheng, W., Hayes, R. L. & Wang, K. K. (2011). Dual vulnerability of tau to calpains and caspase-3 proteolysis under neurotoxic and neurodegenerative conditions. *ASN Neuro.*, *3*, e00051.

[174] Park, S. Y. & Ferreira, A. (2005). The generation of a 17 kDa neurotoxic fragment: an alternative mechanism by which tau mediates β□amyloid-induced neurodegeneration. *J. Neurosci.*, *25*, 5365-5375.

[175] Raynaud, F. & Marcilhac, A. (2006). Implication of calpain in neuronal apoptosis. A possible regulation of Alzheimer's disease. *FEBS J.*, *273*, 3437-3443.

[176] Takahashi, K., Kong, Q., Lin, Y., Stouffer, N., Schulte, D. A., Lai, L., Liu, Q., Chang, L. C., Dominguez, S., Xing, X., Cuny, G. D., Hodgetts, K. J., Glicksman, M. A. & Lin, C. L. (2015). Restored glial glutamate transporter EAAT2 function as a potential therapeutic approach for Alzheimer's disease. *J. Exp. Med.*, *212*, 319-332.

[177] Hamilton, A., Esseltine, J. L., DeVries, R. A., Cregan, S. P. & Ferguson, S. S. (2014). Metabotropic glutamate receptor 5 knockout reduces cognitive impairment and pathogenesis in a mouse model of Alzheimer's disease. *Mol. Brain*, *7*, 40.

[178] Kornhuber, J. & Weller, M. (1997). Psychotogenicity and N-methyl-D-aspartate receptor antagonism: implications for neuroprotective pharmacotherapy. *Biol. Psychiatry*, *41*, 135-144.

[179] Farlow, M. R. (2004). NMDA receptor antagonists. A new therapeutic approach for Alzheimer's disease. *Geriatrics 59*, 22-27.

[180] Sonkusare, S. K., Kaul, C. L. & Ramarao, P. (2005). Dementia of Alzheimer's disease and other neurodegenerative disorders--memantine, a new hope. *Pharmacol. Res.*, *51*, 1-17.

[181] Wenk, G. L., Parsons, C. G. & Danysz, W. (2006). Potential role of N-methyl-D-aspartate receptors as executors of neurodegeneration resulting from diverse insults: focus on memantine. *Behav. Pharmacol.*, *17*, 411-424.

[182] Lipton, S. A. & Chen, H. S. (2005). Paradigm shift in NMDA receptor drug development. *Expert Opin. Ther. Targets*, *9*, 427-429.

[183] Herrmann, N., Li, A. & Lanctot, K. (2011). Memantine in dementia: a review of the current evidence. *Expert. Opin. Pharmacother.*, *12*, 787-800.

[184] Parsons, C. G. & Gilling, K. (2007). Memantine as an example of a fast, voltage-dependent, open channel N-methyl-D-aspartate receptor blocker. *Methods Mol. Biol.*, *403*, 15-36.

[185] Léveillé, F., El Gaamouch, F., Gouix, E., Lecocq, M., Lobner, D., Nicole, O. & Buisson, A. (2008). Neuronal viability is controlled by a functional relation between synaptic and extrasynaptic NMDA receptors. *FASEB J.*, *22*, 4258-4271.

[186] Xia, P., Chen, H. S., Zhang, D. & Lipton, S. A. (2010). Memantine preferentially blocks extrasynaptic over synaptic NMDA receptor currents in hippocampal autapses. *J. Neurosci.*, *30*, 11246-11250.

[187] Rao, V. L., Dogan, A., Todd, K. G., Bowen, K. K. & Dempsey, R. J. (2001). Neuroprotection by memantine, a non-competitive NMDA receptor antagonist after traumatic brain injury in rats. *Brain Res.*, *911*, 96-100.

[188] Lorrio, S., Negredo, P., Roda, J. M., García, A. G. & López, M. G. (2009). Effects of memantine and galantamine given separately or in association, on memory and hippocampal neuronal loss after transient global cerebral ischemia in gerbils. *Brain Res.*, *1254*, 128-137.

[189] Zhao, X., Marszalec, W., Toth, P. T., Huang, J., Yeh, J. Z. & Narahashi, T. (2006). *In vitro* galantamine-memantine co-application: mechanism of beneficial action. *Neuropharmacol.*, *51*, 1181-1191.

[190] Colom, L. V., Castaneda, M. T., Aleman, D. & Touhami, A. (2013). Memantine protects cholinergic and glutamatergic septal neurons from Aβ1-40-induced toxicity. *Neurosci. Lett.*, *541*, 54-57.

[191] Klyubin, I., Wang, Q., Reed, M. N., Irving, E. A., Upton, N., Hofmeister, J., Cleary, J. P., Anwyl, R. & Rowan, M. J. (2011). Protection against Aβ-mediated rapid disruption of synaptic plasticity and memory by memantine. *Neurobiol. Aging*, *32*, 614-623.

[192] Alley, G. M., Bailey, J. A., Chen, D., Ray, B., Puli, L. K., Tanila, H., Banerjee, P. K. & Lahiri, D. K. (2010). Memantine lowers amyloid-beta peptide levels in neuronal cultures and in APP/PS1 transgenic mice. *J. Neurosci. Res.*, *88*, 143-154.

[193] Decker, H., Jürgensen, S., Adrover, M. F., Brito-Moreira, J., Bomfim, T. R., Klein, W. L., Epstein, A. L., De Felice, F. G., Jerusalinsky, D. & Ferreira, S. T. (2010). N-methyl-D-aspartate receptors are required for synaptic targeting of Alzheimer's toxic amyloid-β peptide oligomers. *J. Neurochem.*, *115*, 1520-1529.

[194] Rammes, G., Hasenjager, A., Sroka-Saidi, K., Deussing, J. M. & Parsons, C. G. (2011). Therapeutic significance of NR2B-containing NMDA receptors and mGluR5 metabotropic glutamate receptors in mediating the synaptotoxic effects of β-amyloid oligomers on long-term potentiation (LTP) in murine hippocampal slices. *Neuropharmacology*, *60*, 982-990.

[195] Mota, S. I., Ferreira, I. L. & Rego, A. C. (2014). Dysfunctional synapse in Alzheimer's disease - A focus on NMDA receptors. *Neuropharmacology*, *76*, 16-26.

[196] Martinez-Coria, H., Green, K. N., Billings, L. M., Kitazawa, M., Albrecht, M., Rammes, G., Parsons, C. G., Gupta, S., Banerjee, P. & LaFerla, F. M. (2010). Memantine improves cognition and reduces Alzheimer's-like neuropathology in transgenic mice. *Am. J. Pathol.*, *176*, 870-880.

[197] Minkeviciene, R., Banerjee, P. & Tanila, H. (2004). Memantine improves spatial learning in a transgenic mouse model of Alzheimer's disease. *J. Pharmacol. Exp. Ther.*, *311*, 677-682.

[198] Van Dam, D. & De Deyn, P. P. (2006). Cognitive evaluation of disease-modifying efficacy of galantamine and memantine in the APP23 model. *Eur. Neuropsychopharmacol.*, *16*, 59-69.

[199] Dong, H., Yuede, C. M., Coughlan, C., Lewis, B. & Csernansky, J. G. (2008). Effects of memantine on neuronal structure and conditioned fear in the Tg2576 mouse model of Alzheimer's disease. *Neuropsychopharmacol.*, *33*, 3226-3236.

[200] Reisberg, B., Doody, R., Stoffler, A., Schmitt, F., Ferris, S., Möbius, H. J. & Memantine Study Group. (2003). Memantine in moderate-to-severe Alzheimer's disease. *N. Engl. J. Med.*, *348*, 1333-1341.

[201] van Dyck, C. H., Tariot, P. N., Meyers, B. & Malca Resnick., E., Memantine MEM-MD-01 Study Group. (2007). A 24-week randomized, controlled trial of memantine in patients with moderate-to-severe Alzheimer disease. *Alzheimer Dis. Assoc. Disord.*, *21*, 136-143.

[202] Winblad, B., Jones, R. W., Wirth, Y., Stoffler, A. & Mobius, H. J. (2007). Memantine in moderate to severe Alzheimer's disease: a meta-analysis of randomised clinical trials. *Dement. Geriatr. Cogn. Disord.*, *24*, 20-27.

[203] Parsons, C. G., Danysz, W., Dekundy, A. & Pulte, I. (2013). Memantine and cholinesterase inhibitors: complementary mechanisms in the treatment of Alzheimer's disease. *Neurotox. Res.*, *24*, 358-369.

[204] Atri, A., Shaughnessy, L. W., Locascio, J. J. & Growdon, J. H. (2008). Long-term course and effectiveness of combination therapy in Alzheimer disease. *Alzheimer Dis. Assoc. Disord.*, *22*, 209-221.

[205] Porsteinsson, A. P., Grossberg, G. T., Mintzer, J. & Olin, J. T., Memantine MEM-MD-12 Study Group. (2008). Memantine treatment in patients with mild to moderate Alzheimer's disease already receiving a cholinesterase inhibitor: a randomized, double-blind, placebo-controlled trial. *Curr. Alzheimer Res.*, *5*, 83-89.

[206] Tariot, P. N., Farlow, M. R., Grossberg, G. T., Graham, S. M., McDonald, S., Gergel, I. & Memantine Study Group. (2004). Memantine treatment in patients with moderate to severe Alzheimer disease already receiving donepezil: a randomized controlled trial. *JAMA*, *291*, 317-324.

[207] Lopez, O. L., Becker, J. T., Wahed, A. S., Saxton, J., Sweet, R. A., Wolk, D. A., Klunk, W. & Dekosky, S. T. (2009). Long-term effects of the concomitant use of memantine with cholinesterase inhibition in Alzheimer disease. *J. Neurol. Neurosurg. Psychiatry*, *80*, 600-607.

[208] Atri, A., Molinuevo, J. L., Lemming, O., Wirth, Y., Pulte, I. & Wilkinson, D. (2013). Memantine in patients with Alzheimer's disease receiving donepezil: new analyses of efficacy and safety for combination therapy. *Alzheimers Res. Ther.*, *5*, 6.

[209] Bakchine, S. & Loft, H. (2007). Memantine treatment in patients with mild to moderate Alzheimer's disease: results of a randomised, double-blind, placebo-controlled 6-month study. *J. Alzheimers Dis.*, *11*, 471-479.

[210] Schneider, L. S., Dagerman, K. S., Higgins, J. P. & McShane, R. (2011). Lack of evidence for the efficacy of memantine in mild Alzheimer disease. *Arch. Neurol.*, *68*, 991-998.

[211] Lipton, S. A. (2007). Pathologically activated therapeutics for neuroprotection. *Nat. Rev. Neurosci.*, *8*, 803-808.

[212] Sozio, P., Cerasa, L. S., Laserra, S., Cacciatore, I., Cornacchia, C., Di Filippo, E. S., Fulle, S., Fontana, A., Di Crescenzo, A., Grilli, M., Marchi, M. & Di Stefano, A. (2013). Memantine-sulfur containing antioxidant conjugates as potential prodrugs to improve the treatment of Alzheimer's disease. *Eur. J. Pharm. Sci. 49*, 187-198.

In: Modulators of Glutamatergic Signaling
Editor: Zoran M. Pavlovic

ISBN: 978-1-63483-493-3

Chapter 7

NEUROTRANSMISSION IN SCHIZOPHRENIA

*Marc Debelle**

ABSTRACT

Dopamine theory of schizophrenia does not account for all the clinical features of schizophrenia. Glutamate Hypothesis gives a holistic explanation of the symptoms of schizophrenia since the blockade of the NMDA receptors induce not only psychotomimetic symptoms, but negative symptoms and cognitive deficits symptoms similar to those present schizophrenia. So far, new chemical entities failed to deliver clinical promises when tested in adequate and well-controlled clinical studies. Much hope is based upon modulators of glutamatergic neurotransmission, since they may be less prone to receptors desensitization.

Keywords: schizophrenia, glutamate, NMDA receptors, modulators

DEFINITION AND CLINICAL FEATURES OF SCHIZOPHRENIA

Schizophrenia is the most common diagnosed psychotic disorder. The incidence and prevalence of schizophrenia are the same all over the world. The prevalence of schizophrenia is related to age, increasing until the age of 40 after which it declines in both sexes. Age of onset is in the late teens in men and 3-5 years later in women, but the overall prevalence of schizophrenia is equal in both sexes.

Clinical symptoms of schizophrenia encompass a wide range of dysfunctions, which are grouped in dimensions. Positive symptoms arean excess or distortion of normal functions such as: delusions (often persecutory, referential or somatic and bizarre in nature); hallucinations (most often auditory in the form of voices conversing or commenting the patient); disorganized speech (incoherence, irrelevant speech, neologisms); and behavioral disturbances (e.g., difficulties in activities of daily living, apparently unmotivated agitation, catatonic behavior) [1]. Negative symptoms reflect a reduction or loss of normal functions.

* Corresponding author: Email: debellemarc68@gmail.com.

Negative symptoms include blunting of affect, poverty of speech, inability to pursue goal-directed activities, anhedonia and social withdrawal [1]. They are often more evident during periods when symptoms associated with acute exacerbations, which are usually positive symptoms, are not at the forefront of the clinical picture. Depressive symptoms and extra pyramidal adverse reactions resulting from antipsychotic treatment can mimic negative symptoms of schizophrenia.

Symptoms of depression frequently occur in patients with schizophrenia. Various studies estimate rates over the longitudinal course of schizophrenia from 7 to 65% [2]. Depressive symptoms are associated with poor quality of life, poorer prognosis and increased risk of suicide.

During the last decades, it has become more evident that schizophrenia is associated with deficits of cognitive functions, impairing community functioning and outcome measures [3, 4, 5]. The cognitive deficits are present at the onset of illness, persist for most of the patient's life without remission, and may precede the development of psychotic symptoms [6]. Beneficial effect on cognitive deficits may have high impact on patient's quality of life.

In addition to the classic symptoms above, schizophrenia is also characterized by primary metabolic disturbances and primary motoric symptoms. Metabolic disturbances, elevated blood glucose and lipids levels and increase in weight, and abnormal movements, parkinsonian symptoms, akathisia, and dyskinesias are known adverse reactions from antipsychotic medications; however, increased occurrence of diabetes and cardiovascular disorder, as well asdisturbances in gait, movements and EPS also occur in untreated patients.

Diagnostic criteria of schizophrenia, according to Diagnostic and Statistical Manual of Mental Disorders (DSM)-V, require two or more specified symptoms of schizophrenia. Criterion A lists five key symptoms of psychotic disorders: 1) delusions, 2) hallucinations, 3) disorganized speech, 4) disorganized or catatonic behaviour, and 5) negative symptoms. In DSM-V, two of these five symptoms are required and at least one symptom must be one of the first three (delusions, hallucinations, disorganized speech, considered as core positive symptoms). The duration of those symptoms should be at least 1 month (or less if their treatment is successful). Criterion B states the functional impairment associated with the disorder, since the disturbance must be associated with a lower level of functioning (work, interpersonal relations, or self-care) markedly below the level achieved prior to the onset of the symptoms. Signs of the disturbance must be present for at least 6 months (Criterion C). During that 6-month period, the symptoms of Criterion A must be present for at least 1 month and may be residual or prodromal before or after meeting Criterion A. During the prodromal or residual period, the only symptoms may be negative symptoms or core attenuated positive symptoms, like odd beliefs, unusual perceptual experiences. Must be ruled out: manic symptoms (to exclude bipolar disorder with psychotic features); major depressive episode (to exclude schizoaffective disorder). When depressive symptoms are present, they must be short in duration compared to the key symptoms seen in the acute exacerbations or residual periods (Criterion D). Similarly, substance use or abuse or a medical condition which could mimic the key symptoms must also be ruled out (Criterion E). If there is a history of autism or communication disorder of childhood onset, schizophrenia diagnosis can only be made in presence of prominent delusions or hallucinations for at least 1 month or less if successfully treated (Criterion F).

Those criteria have a wide diagnostic validity. Eighty to 90 percent of the diagnosed patients are always diagnosed as with schizophrenia after 1 to 10 years [7]. The concordant validity (e.g., by biomarkers) is much lower. Of note, cognitive dysfunction is not a key symptom or criterion to establish a schizophrenia diagnosis. Cognitive deficits are not sufficiently discriminative between schizophrenia and other mental disorders. However, the persistence and aggravation of cognitive deficits may contribute to the diagnosis and the differentiation between schizophrenia and other conditions [8].

The other diagnostic classification used in medicine, the International Classification of Disease, has undertaken a significative lifting. ICD 11 is not yet finalised. However, it seems that the 2 classifications will be better harmonised than earlier [9]. DSM-V alterations from the DSM-IV will likely be included in ICD-11 as well. The definition of schizophrenia sub-types will not be included. Instead, schizophrenia dimensions will be added. Moreover, the peculiar treatment importance of first-rank symptoms according to Schneiderian definition will no longer be kept in both classifications. However, some discrepancies will persist between both classifications. They are related to the duration of the symptoms: 6 months in the DSM versus 1 month in the ICD; also the requirement for a functional disability present in DSM will still be absent in ICD.

One of the most important changebetween DSM V and IV is the focus given on schizophrenia dimensions rather than sub-types. This change may benefit in the future to new therapeutic approaches targeting negative symptoms and cognitive dysfunctions. The current holistic view of schizophrenia leads to the stabilisation of positive symptoms prior to conduct clinical trials with co-treatments to antispychotics in order to focus on non-psychotic symptoms. The dimensional approach of schizophrenia may impact the clinical development of new treatment in schizophrenia patients with predominant negative symptoms or cognitive dysfunctions.

There are potential overlaps between the schizophrenia dimensions. Measures of cognitive functioning are more likely to be associated with the severity of negative symptoms than with the severity of positive symptoms [10].

Clinically there is an overlap in the definitions of cognitive and negative symptoms. Some clinical rating systems for negative symptoms consider deficits in abstract thinking and stereotyped thinking, clearly cognitive processes, to be negative symptoms (e.g., PANSS). In such a rating system, the overall severity of negative symptoms may very well correlate with cognitive performance. Therefore, the PANSS factor score for negative symptoms is considered as a more reliable measurement of negative symptoms, with less interference from cognitive dysfunction. SANS consider functional deficits to be negative symptoms, leading to high correlation between cognitive and negative symptoms. A gradient of correlation between negative and cognitive symptoms has been suggested. Blunted affect is least strongly correlated with cognitive performance, while alogia, social, and occupational deficits are more strongly correlated (Harvey et al. 1997). We may distinguish negative symptoms and cognitive symptoms in other ways. It has been demonstrated that the severity of cognitive and negative symptoms does not show the patterns of correlation over time that would suggest a causal relationship [11]. The nature of the relationship between cognitive and negative symptoms, and the ability to clinically separate these domains in efficacy studies with compounds aimed at treating negative and/or cognitive symptoms, remains to be clarified.

NEUROTRANSMITTER HYPOTHESIS OF SCHIZOPHRENIA

Dopamine Theory of Schizophrenia

The clinical efficacy of neuroleptics, as shown in the years 1950 – 60 [12, 13], has been the starting point of the dopaminergic theory of schizophrenia. Further to an administration of chlorpromazine to animals, an increase is followed by a decrease of the dopamine metabolites in the brain. Similar, dopamine agonists like amphetamines induced psychotic symptoms mimicking those of schizophrenia. Neuroleptics were then identified as Dopamine D2 receptor antagonists [14]. The hypothesis of a hyperdopaminergic activity in patients was the main biologic explanation of schizophrenia. However, new concepts of the dopaminergic state in schizophrenia have progressed, to better model the dopaminergic disturbances of schizophrenia. It is currently thought there is a dopaminergic imbalance in schizophrenia [15]. The sub-cortical hyperdopaminergic state is related to positive symptoms, while ahypodopaminergic state is found in the prefrontal cortex and would be responsible for negative symptoms of schizophrenia.

Grace showed also that dopaminergic neurons are activated according to 2 modes:

1. A regular and slow tonic firing
2. A peak phasic tone.

Therefore, he suggested the negative symptoms of schizophrenia may be related to hypofrontality state, leading to a decrease of glutamic acid release in the sub-cortical regions of the brain. This low release would induce the hyperactivity of the mesolimbic dopaminergic neurons located in the mesolimbic pathway. In turn, those neurons would become insensitive to external stimuli. This theory is consistent with the clinical observation that in adolescents negative symptoms precede the occurrence of psychotic symptoms and are the first symptoms of schizophrenia, often retrospectively identified as part of the disease.

In light of this theory, the hypothesis of dopaminergic state in schizophrenia is not sufficient to account for the biological explanation of the disease and the whole clinical symptoms, in particular if considered in dimensions. An interaction between dopamine and glutamine seem to be present and be responsible for the different types of symptoms.

Glutamate Hypothesis of Schizophrenia

In the late 1950s, occurrences of psychoses were observed secondary to the intake of phencyclidine (PCP), sometimes as early as secondary to the first PCP intake. The clinical picture consisted of prolonged schizophreniform symptoms which lasted on average 2 to 4 weeks. Contrary to the amphetamine-induced psychotic symptoms, PCP and ketamine (PCP-derivative) induced psychotic symptoms as well as negative symptoms and cognitive dysfunctions. Both compounds are N-methyl-D-aspartate (NMDA) receptor antagonist. They block the NMDAR channel [16].

The absolute concentrations and the potency with which several compounds induce psychosis in humans and animal models is related to the rank order of potency at the

NMDAR, but not at other receptor types [16, 17]. All compounds binding to the NMDAR have been found to induce psychotic symptoms in human, an unique feature related to receptor binding. Therefore, NMDAR blockade appears to be a necessary and sufficient condition to explain psychotomimetic effects. Those effects are not mediated only via the PCP-site. NMDAR antagonists acting via other sites than the PCP site, such as the glutamate binding site and the glycine modulatory site, also have psychotomimetic effects [18, 19].

The glutamate theory of schizophrenia also reconsiders the brain regions involved in schizophrenia. Dopamine projections in the brain are relatively discrete and a limited number of brain regions seems to dysfunction, according to the dopamine theory. Specifically, symptoms of schizophrenia would be due to dysfunctions in the dorso-lateral pre-frontal cortex [20] and the striatum [21]. Glutamate transmission is distributed throughout cortical and subcortical brain regions [24]. As an example, mismatch negativity (MMN), an event-related potential indexing brain function at the level of auditory cortex, has been shown to be abnormal in patients with schizophrenia [24], in normal subjects exposed to ketamine [22, 23], and in awake primates who received NMDAR antagonists into auditory cortex [24].

How leads NMDAR hypofunction to schizophrenia symptoms?

There are 2 broad categories of presynaptic and postsynaptic hypotheses [25]

Presynaptic Hypotheses

The NMDAR hypofunction does not lead to low levels of CSF glutamate level [17], suggesting that more complex disturbances in glutamatergic function might occur. In the contrary, it has been demonstrated that hyper, rather than hypo, glutamatergic function might be evolved. This hyperfunction may be mediated through activation of AMPA receptors.

In awake animals, systemic injections of NMDAR antagonists producing cognitive dysfunction and motor stereotypy increased glutamergic efflux in the prefrontal cortex [26, 27]. However, the blockade of AMPAR reduced the motoric and cognitive dysfunctions induced by NMDAR blockade [28]. Thus, NMDAR antagonists seem to increase glutamate release at some synapses, which then abnormally increase glutamate neurotransmission at non NMDAR, in particular at AMPAR. This finding suggests that psychotomimetic consequences of NMDAR deficiency is due to dysregulation of glutamate neurotransmission, involving NMDAR hypofunction and non-NMDAR hyperfunction.

It is currently postulated that this mechanism involves the so-called cortical dishinbition process [25]. The activity of hippocampal and neocortical pyramidal neurons is under the control of GABA interneurons. Without GABAergic inhibition, excitatory inputs on pyramidal cells would cause a chain reaction of ever increasing activation. The regulation of the firing of pyramidal neurons by GABA interneurons is critical for the coordination of the cortical-mediated behaviours [29]. Increased discharge of GABAergic interneurons results in decreased discharge of pyramidal neurons, and the contrary. In the neocortex and hippocampus, interneurons have a lower threshold for activation compared with pyramidal cells [30]. NMDAR channels, which are voltage-gated channels, are open on those neurons. Thus when exposed to an NMDAR antagonist, GABA neurons will preferentially be inhibited and pyramidal cells will artificially be disinhibited (no more control is performed on those by inhibited interneurons). This dishinbition will lead to an artificial engagement of pyramidal cells or noise which may be recorded in awake animals [31], and an enhanced glutamate release from local pyramidal cells, which will excessively activate AMPAR [28].

Postsynaptic Hypothesis

NMDARs are composed of variable combinations of subunits. Complex alterations in NMDAR subunit composition have been identified in schizophrenia, along with alterations in specific NMDAR-related postsynaptic proteins [32]. NMDARs are localized at postsynaptic dendritic terminals, as well as on presyaptic terminals where they control glutamate release (see above), and also on oligodendrocytes [33]. The exact role of specific NMDAR subunits remains to be clarified.

Several disturbances in modulatory symptoms have also been demonstrated in schizophrenia. They include disturbances in the synthesis and catabolism of glycine and D-serine, which bind to the glycine modulatory site of the NMDAR [25], and glutathione, which acts on the redox site [34].

Glycine is synthethised from L-serinby one enzyme and regulated in the synapse by the glycine T-1 transporter [25].

D-serine is synthethised in the brain from L-serine by one specific enzyme, and catabolised by another enzyme. Genetic studies and animal studies have shown associations of both enzymes with schizophrenia. Reductions in D-serine levels in CSF have been demonstrated in patients with schizophrenia [35].

Glutatione levels have been demonstrated to be reduced in patients with schizophrenia (25Many factors may affect glutathione levels and witnesses nonspecific brain injury mechanisms. N-acetylcysteine [36], a dietary glutathione precursor, may modulate to some extent brain glutathione levels.

Allosteric Modulatory Sites

Traditional agonists or antagonists of glutamate neurotransmission at pre- or postsynaptic receptors may lead to neurotoxicity or responses which may be detrimental to the cortical function, in case of sustained activation of the NMDAR. Allosteric modulatory sites enhance or reduce the function in response to incoming stimulus, in an activity dependent manner. They only work to modulate the function of the receptor when the receptor is stimulated by its natural neurotransmitter. Two sets of allosteric modulatory sites exist [25]. Those inherent to the NMDAR itself and those inherent to receptors that may serve to modulate NMDAR-mediated neurotransmission.

Potential Treatments of Schizophrenia Based upon Glutamate Neurotransmission

Metabotropic Glutamate Group 2/3 Receptor Agonists

An example of those compounds is the orthosteric agonist, LY2140023 or Pomaglumetadmethionil. It was expected that metabotropic glutamate(mGlu) receptor 2/3 agonists acting at pyramidal cells may restore normalized cortical activity, disturbed by hypofunctioning of NMDAR located on GABAergic interneurons and restore the function of cortical pyramidal cells and thus normalize the behavioral abnormalities.

The POC study [37] showed greater efficacy of LY2140023 than placebo on positive and negative symptoms of schizophrenia. Olanzapine included as the third treatment arm validated the assay. The second study [38] did not give indications for an effective dose of the compound. This dose-finding study was not conclusive, since any of the treatment arm (LY2140023 and olanzapine) failed to show significant efficacy benefit over placebo. The results from a long-term safety study demonstrated that LY2140023was comparable to standard of care (SOC) therapy in time to discontinuationdue to adverse events (AEs) [38]. However, more patients discontinued LY2140023 than olanzapine discontinued the study due to lack of efficacy. The fourth and currently last study [39] found no benefit of adjunctive LY2140023 versus placebo for negative symptoms in patients with schizophrenia receiving treatment with standard of care. LY2140023 was generally well-tolerated in these patients. Of note, continued use of orthosteric agonist may lead to receptor desensitization and thus loss of efficacy.

Compounds Acting on Intrinsic Sites

Endogeneous Ligands

The most investigated intrinsic site has been the glycine modulatory site, because endogenous ligands, glycine and D-serine, are natural substances and can be used in clinical trials. Significant effects have been observed in single-site studies [40, 41, 42, 43] but failed to replicate in multicenter studies [25]. In absence of an active comparator for assay validity, no conclusion can be drawn about the efficacy of the tested compounds. Moreover, dose issues have been reported in those trials and the receptor occupancy at the investigated doses was unknown. Therefore the relevance of the tested doses to adequately assess the efficacy is still an open question.

Gly T1 Inhibition

Blocking the function of Glycine T1 transporters is based on the same concept than for serotonine in depression, with the aim to obtain a natural glycine increase in the synaptic cleft [25]. Several glycine transporter inhibitors have been studied in preclinical models of schizophrenia and they were able to reverse PCP-related hyperactivity with parallel rank order of potency to their effects on glycine transport. One such compound is sarcosine which was effective in multiple animal models of schizophrenia; however showed motor and pulmonary toxicity in in vivo animal studies. It seems however that these side effects are not related to NMDAR. In humans, sarcosine has been effective in acute and chronic schizophrenia in small-scale studies [44, 44, 45, 46]. However, similar as for glycine, it is unknown whether the tested doses are clinically optimal doses due to lack of biomarker able to engage on the glycine binding-site.

Recently, the first selective, high-affinity GlyT1 compound bitopertin was studied in a clinical development plan in patients with negative symptoms of schizophrenia or suboptimally treated. Dose tested were based upon a PET study of glycine-site occupancy, with dose chosen to prevent activation-related MDAR desensitization [25, 47]. A 2014 press release from the Sponsor informed of the failure to demonstrate higher efficacy from bitopertin compared to placebo when added on to antipsychotic treatments in patients with predominant negative symptoms.

DAAO Inhibition

DAAO is the enzyme which catabolizes D-serine. Combining D-serine to DAAO may prevent renal and brain D-serine degradation, and increases significantly D-serine potency in animal models [48]. A few clinical trials in patients receiving add-on DAAO inhibitors to antipsychotic are registered in ClinicalTrials.gov (NCT02261519, NCT00960219, NCT02026947). They are either completed unreported, or ongoing. Therefore, ultimate utility of this approach is still to be determined.

Compounds Acting at Extrinsic Sites

Compounds targeting metabotropic glutamate receptors may modulate glutamate presynaptically or modulate NMDAR postsynaptically. mGlu2/3 receptors modulate and limit glutamate release. Full m2/3Glu agonists carry the risk of receptor downregulation over time with development of tolerance and subsequent loss of efficacy. Therefore ositive allosteric modulators at this site may be more suitable [49].

Another potential target is mGlu5 receptors at postsynaptic level. They are located at some synapses close to NMDAR and modulate the dynamics of NMDAR channels. A number of mGluR5 agonists and PAM has been synthethised and shown to be promising in preclinical models of schizophrenia. However, none of them have entered into clinical development in schizophrenia.

Compounds Acting on AMPA-Type Glutamate Receptors and Other Targets

Agents stimulating AMPA-type glutamate receptors have not shown efficacy for improving symptoms of schizophrenia or cognitive dysfunction associated with the condition [49]. The clinical development of xanomeline, a M1/M4 muscarinic agent, has been discontinued to gastrointestinal side effects.

CONCLUSION

The glutamatergic theories of schizophrenia account for a holistic approach of schizophrenia symptoms, based not only of positive symptoms but also on negative symptoms and cognitive dysfunction. Future directions will go towards the development of reliable biomarkers, which will help to successfully develop compounds acting on and modulating glutamate neurotransmission. Those biomarkers should help to adequately delineating the doses to be tested and to prevent the development of tolerance, which so far have prevented the compounds based upon glutamate neurotransmitters to demonstrate clinical evidence in the treatment of patients with schizophrenia.

REFERENCES

[1] Freedman, R. (2003). Schizophrenia, *N Engl J Med*, *349*, 1738-49.

[2] Siris, S. G. (1995). Depression and schizophrenia. In Schizophrenia (eds S. R. Hirsch & D. R. Weinberger), 128-145. Oxford: Blackwell.

[3] Harvey, P. D. & Keefe, R. S. (2001). Studies of cognitive change in patients with schizophrenia following novel antipsychotic treatment. *Am J Psychiatry*, Feb, *158*(2), 176-84.
[4] Green, M. F., Kern, R. S., Braff, D. L., et al. (2000). Neurocognitive deficits and functional outcome in schizophrenia: Are we measuring the "right stuff? *Schizophr Bull*, *26*, 119–36.
[5] Sharma, T. & Antonova, L. (2003). Cognitive function in schizophrenia. Deficits, functional consequences, and future treatment. *Psychiatr Clin North Am.*, *26*(1), 25-40.
[6] Tollefson, G. D. & Sanger, T. M. (1997). Negative symptoms: a path analytic approach to a double-blind, placebo- and haloperidol-controlled clinical trial with olanzapine. *Am J Psychiatry*, *154*(4), 466–474.
[7] Bromet, E. J., Kotov, R., Fochtmann, L. J., et al. (2011). Diagnostic shifts during the decade Following first admission for psychosis. *Am. J. Psychiatry*, *168*, 11861194.
[8] Tandon, R., et al. (2013). Definition and description of schizophrenia in the DSM-5. Schizophr. Res
[9] Gaebel W, (2012). Status of Psychotic Disorders in ICD-11. *Schizophr Bull*, (2012) *38* (5), 895-898.
[10] Addington, J. (2000). Negative symptoms and the relationship to cognitive functioning in Sharma, T., & Harvey, P. (Eds.), *Cognition in schizophrenia.* Oxford, United Kingdom: Oxford University Press.
[11] Harvey, P. D. & Keefe, R. S. E. (1997). Cognitive impairment in schizophrenia and implications of atypical neuroleptic treatment. *CNS Spectrums*, *2*, 1–11.
[12] Delay et Deniker, 38 cas de psychoses traitées par la cure prolongée et continue de 4560 RP, comptes-rendus du Congrès d'aliénation et de neurologie de langue française, Paris, Masson, 1952, p. 497-502.
[13] Carlsson, A. & Lindqvist, M. (1963). Effect of chlorpromazine or haloperidol on formation of 3-methoxytyramine and normetanephrine in mouse brain. *Acta Pharmacol Toxicol (Copenh)*, *20*,140–144.
[14] Seeman, P., et al. (1976). *Antipsychotic drug doses and neuroleptic/dopamine receptors Nature*, *261*, 717 – 719.
[15] Weinberger, D. R. (1987). Implications of normal brain development for the pathogenesis of schizophrenia. *Archives of general psychiatry*, *44*(7), 660-669.
[16] Javitt, D. & Zukin, S. R. (1991). Recent advances in the phencyclidine model of schizophrenia. *Am J Psychiatry*, *148*(10):1301-8.
[17] Seeman, P. (2010). All roads to schizophrenia lead to dopamine supersensitivity and elevated dopamine D2(high) receptors. *CNS Neuroscience and therapeutics*, *17*(2), 118-32.
[18] Muir, J., Everitt, B., et al. (1995). Reversal of visual attentional dysfunction following lesions of the cholinergic basal forebrain by physostigmine and nicotine but not by the 5-HT3 receptor antagonist, ondansetron. *Psychopharmacology.*,*118*, 82–92.
[19] Preskorn, S. H., Baker, B., et al. (2008). An innovative design to establish proof of concept of the antidepressant effects of the NR2B subunit selective N-methyl-D-aspartate antagonist, CP-101,606, in patients with treatment-refractory major depressive disorder. *J Clin Psychopharmacol.*, *28*, 631–637.
[20] Lesh, T. A., Niendam, T. A., et al. (2011). Cognitive control deficits in schizophrenia: mechanisms and meaning. *Neuropsychopharmacol.*, 36, 316–338.

[21] Simpson, E. H., Kellendonk, C., et al. (2010). A possible role for the striatum in the pathogenesis of the cognitive symptoms of schizophrenia. *Neuron.*, *65*, 585–596.

[22] Krystal, J. H., Karper, L. P., et al. (1994). Subanesthetic effects of the noncompetitive NMDA antagonist, ketamine, in humans. Psychotomimetic, perceptual, cognitive, and neuroendocrine responses. *Arch Gen Psychiatry.*, *51*, 199–214.

[23] Heekeren, K., Daumann, J., et al. (2008). Mismatch negativity generation in the human 5HT2A agonist and NMDA antagonist model of psychosis. *Psychopharmacology.*, 199, 77–88.

[24] Javitt, D. C., Steinschneider, M., et al. (1996). Role of cortical N-methyl-D-aspartate receptors in auditory sensory memory and mismatch negativity generation: implications for schizophrenia. *Proc Natl Acad Sci USA.*, *93*,11962–11967.

[25] Moghaddam, B. & Javitt, D. (2012). From Revolution to Evolution: The Glutamate Hypothesis of Schizophrenia and its Implication for Treatment. *Neuropsychopharmacology*, *37*(1), 4–15.

[26] Liu, J. & Moghaddam, B. (1995). Regulation of glutamate efflux by excitatory amino acid receptors: evidence for tonic inhibitory and phasic excitatory regulation. *J Pharmacol Exp Ther.*

[27] Lorrain, D. S. & Baccei, C. S., et al. (2003). Effects of ketamine and N-methyl-D-aspartate on glutamate and dopamine release in the rat prefrontal cortex: modulation by a group II| selective metabotropic glutamate receptor agonist LY379268. *Neuroscience.*

[28] Moghaddam, B., Adams, B., et al. (1997). Activation of glutamatergic neurotransmission by ketamine: a novel step in the pathway from NMDA receptor blockade to dopaminergic and cognitive disruptions associated with the prefrontal cortex. *J Neurosci.*,*17*, 2921–2927.

[29] Buzsaki, G., Geisler, C., et al. (2004). Interneuron diversity series: circuit complexity and axon wiring economy of cortical interneurons. *Trends Neurosci.*, 27, 186–193.

[30] Csicsvari, J., Hirase, H., et al. (1998). Reliability and state dependence of pyramidal cell-interneuron synapses in the hippocampus: an ensemble approach in the behaving rat. *Neuron.*, *21*, 179–189.

[31] Jackson, M., Homayoun, H., et al. (2004). *NMDA receptor hypofunction produces concomitant firing rate potentiation and burst activity reduction in the prefrontal cortex Proc Natl Acad Sci USA*, 1016391–6396, 6396.

[32] Kristiansen, L. V., Huerta, I., et al. (2007). NMDA receptors and schizophrenia. *Curr Opin Pharmacol.*, *7*, 48–55.

[33] Matute, C., Melone, M., et al. (2005). Increased expression of the astrocytic glutamate transporter GLT-1 in the prefrontal cortex of schizophrenics. *Glia*, *49*, 451–455.

[34] Gysin, R., Kraftsik, R., et al. (2007). Impaired glutathione synthesis in schizophrenia: convergent genetic and functional evidence. *Proc Natl Acad Sci USA.*, *104*, 16621–16626.

[35] Labrie, V. & Roder, J. C. (2009). *The involvement of the NMDA receptor D-serine/glycine site in the pathophysiology and treatment of schizophrenia Neurosci Biobehav Rev*, 34351–372.372.

[36] Dodd, S., Dean, O., et al. (2008). N-acetylcysteine for antioxidant therapy: pharmacology and clinical utility. *Expert Opin Biol Ther.*, *8*,1955–1962.

[37] Patil, S. T., Zhang, L., et al. (2007). Activation of mGlu2/3 receptors as a new approach to treat schizophrenia: a randomized Phase 2 clinical trial. *Nat Med.*, *13*, 1102–1107.

[38] Kinon, B. J., Zhang, L., et al. (2011). A multicenter, inpatient, phase 2, double-blind, placebo-controlled dose-ranging study of LY2140023 monohydrate in patients with DSM-IV schizophrenia. *J Clin Psychopharmacol.*, *31*, 349–355.

[39] Stauffer, V. L., et al. (2013). Pomaglumetad methionil: no significant difference as an adjunctive treatment for patients with prominent negative symptoms of schizophrenia compared to placebo. *Schizophr Res.*, *150*(2-3):434-41.

[40] Javitt, D. C., Zylberman, I., et al. (1994). *Amelioration of negative symptoms in schizophrenia by glycine Am J Psychiatry*, 1511234–1236, 1236.

[41] Heresco-Levy, U., Javitt, D. C., et al. (1999). Efficacy of high-dose glycine in the treatment of enduring negative symptoms of schizophrenia. *Arch Gen Psychiatry*, *56*, 29–36.

[42] Tuominen, H., Tiihonen, J., et al. (2005). Glutamatergic drugs for schizophrenia: a systematic review and meta-analysis. *Schizophr Res.*, *72*, 225–234.

[43] Tsai, G. E. & Lin, P. Y. (2010). Strategies to enhance N-methyl-D-aspartate receptor-mediated neurotransmission in schizophrenia, a critical review and meta-analysis. *Curr Pharm Des.*,*16*, 522–537.

[44] Lane, H. Y., Chang, Y. C., et al. (2005). Sarcosine or D-serine add-on treatment for acute exacerbation of schizophrenia: a randomized, double-blind, placebo-controlled study. *Arch Gen Psychiatry.*, *62*, 1196–1204.

[45] Lane, H. Y., Lin, C. H., et al. (2010). A randomized, double-blind, placebo-controlled comparison study of sarcosine (N-methylglycine) and D-serine add-on treatment for schizophrenia. *Int J Neuropsychopharmacol.*, *13*, 451–460.

[46] Lane, H. Y., Liu, Y. C., et al. (2008). Sarcosine (N-methylglycine) treatment for acute schizophrenia: a randomized, double-blind study. *Biol Psychiatry.*, *63*, 9–12.

[47] Tsai, G., Lane, H. Y., et al. (2004). Glycine transporter I inhibitor, N-methylglycine (sarcosine), *added to antipsychotics for the treatment of schizophrenia Biol Psychiatry*, 55452–456, 456.

[48] Umbricht, D., Schmid, L., et al. (2000). Ketamine-induced deficits in auditory and visual context-dependent processing in healthy volunteers: implications for models of cognitive deficits in schizophrenia. *Arch Gen Psychiatry.*, *57*, 1139–1147.

[49] Hashimoto, K., Fujita, Y., et al. (2009). Co-administration of a D-amino acid oxidase inhibitor potentiates the efficacy of D-serine in attenuating prepulse inhibition deficits after administration of dizocilpine. *Biol Psychiatry.*, *65*, 1103–1106.

[50] Conn, P. J., Lindsley, C. W., et al. (2009). *Activation of metabotropic glutamate receptors as a novel approach for the treatment of schizophrenia Trends Pharmacol Sci*, 3025–31, 31.

INDEX

#

A

B

C

D

E

F

G

H

I

J

K

L

M

N

O

P

Q

R

S

T

U

V

W

X

Y